INDUSTRIAL
DECISION-MAKING
AND HIGH-RISK
TECHNOLOGY

INDUSTRIAL DECISION-MAKING AND HIGH-RISK TECHNOLOGY

Siting Nuclear Power Facilities in the USSR

CHARLES K. DODD

ROWMAN & LITTLEFIELD PUBLISHERS, INC.

ROWMAN & LITTLEFIELD PUBLISHERS, INC.

Published in the United States of America
by Rowman & Littlefield Publishers, Inc.
4720 Boston Way, Lanham, Maryland 20706

3 Henrietta Street, London WC2E 8LU, England

British Cataloging in Publication Information Available

Library of Congress Cataloging-in-Publication Data

Dodd, Charles K.
Industrial decision-making and high-risk technology : siting nuclear
power facilities in the USSR / Charles K. Dodd.
p. cm.
Includes bibliographical references.
1. Nuclear power plants—Former Soviet republics—Location.
2. Nuclear power plants—Government policy—Former Soviet republics.
3. Decision-making—Former Soviet republics. I. Title.
TK1362.S65D64 1993 363.17'99'0947—dc20 93–20253 CIP

ISBN 0–8476–7847–4 (cloth : alk. paper)

$$TK1362$$
$$S65$$
$$D64$$
$$1994 \times$$

Printed in the United States of America

∞ ™ The paper used in this publication meets the minimum requirements of
American National Standard for Information Sciences—Permanence of
Paper for Printed Library Materials, ANSI Z39.48–1984.

Contents

OCT 0 4 1994

List of Figures

List of Maps

List of Tables

Acknowledgments

This book would not be possible without the assistance of a number of individuals. I would like to acknowledge the contributions of Judith Thorton, Craig ZumBrunnen and Philip Pryde, Gabriel Gallardo, Ian Sterling, and Delia Rosenblatt who at one time or another provided helpful assistance in their comments on the manuscript. The late Harold Swayze provided invaluable help on translations of many Russian language documents. I would also like to extend thanks to Marcus Lester and Nichole Devine for their assistance on map production. Lastly I would like to acknowledge the efforts of Lynn Gemmell and Beth Mitchell for their patience and diligence in the editing of the manuscript.

Guide to Abbreviations in Source Notes and Bibliography

The following abbreviations are used in the source notes and bibliography:

BBC, SWB—BBC Monitoring, Summary of World Broadcasts
CDPSP—Current Digest of the Post-Soviet Press
CDSP—Current Digest of Soviet Press
JPRS—Joint Publications Research Service
RFE/RL—Radio Free Europe/Radio Liberty
TsSU—Tsentral'noye Statisticheskoye Upravleniye (Central Statistical Administration)

Introduction

The rapid growth of increasingly complex technologies during the 20th century has presented society an entirely new set of problems with social, political, and economic ramifications. One problem area that has been recognized over the past quarter-century concerns public safety and high-risk technologies. High-risk technologies are characterized by low probabilities of potential accidents but with very damaging or lethal consequences resulting from such accidents. One of the most controversial of these high-risk technologies is the commercial application of nuclear power as an energy source. Since 1954, when the first reactor was used to generate electric power for civilian consumption, nuclear power has come to supply more than 17% of global electricity production.[1] Nuclear power is now utilized in at least twenty-five countries for electric power production and, to a much more limited extent, for the generation of steam and hot water.[2] Nevertheless, the widespread adoption of nuclear power as a source of energy has generated considerable debate in many countries.

Although nuclear power is, in many cases, an attractive source of energy for national policy-makers and energy industry officials, there has been considerable disagreement over the desirability of this technology. This controversy stems from the negative externalities or impacts that various segments of society associate with nuclear power technology, particularly with respect to public safety and environmental contamination.

There are several issue areas within the general topic of nuclear power and public safety. These include (1) the reliability and safety of reactor and facility designs; (2) the competence and training of reactor operators and managers; (3) nuclear waste storage and disposal; and (4) nuclear reactor facility siting. The last issue area, siting, attracts considerable interest because plant location is easily perceived and has a more direct impact on individuals. As the nuclear industry has grown and experience has been accumulated, interest and concern over site decisions have increased. In recent years, in many countries that have adopted nuclear power technology, different segments of society concerned over the potential negative impacts of siting decisions have attempted to become involved in the site decision-making process. As a consequence

1

controversies over siting have deepened and expanded. Many countries have established institutions and procedures to facilitate decision-making and produce more acceptable siting solutions for broader sections of society. A growing body of literature suggests that the organization of decision-making, in particular the procedures and composition of participating institutions and social groups, profoundly influences the outcomes of siting decisions as well as the growth and spatial pattern of the technology's use.[3]

One country that has gone through major changes in how siting and technology decisions are made is the former Soviet Union. Changes have occurred in decision-making processes, participants and policies. Since the Chernobyl' accident in 1986, public opposition to nuclear power has arisen throughout the former USSR. Much of the opposition to nuclear power centers on the issue of nuclear power plant siting. For the newly independent states of the former Soviet Union, this issue is of some significance, as the region faces acute regional energy shortages and suffers from an economic system and geographic pattern of resource endowments that have incurred significant societal costs in the use of alternative types of energy.

Nature and Purpose of the Book

In this study I examine the issue of siting commercial nuclear power facilities in the former Soviet Union. While much has been written on the state of affairs in the Soviet nuclear power industry, little has focused specifically on the issues of facility siting.[4] Conversely, while there is a growing body of literature on high-risk technology and site decision-making, it is limited to the experience of a few Western countries.[5] Thus, it is hoped that this study contributes to a better understanding of how siting decisions concerning high-risk technology have been made in the Soviet Union, a country with a completely different political economy than those found in Western countries. Moreover, understanding past decision-making and policy in the Soviet nuclear power industry has important implications for the present as many of the former Soviet institutions and administrative structures are still largely in place. This study is not concerned specifically with site selection methodology; indeed, information on site selection methodology in the former Soviet Union is scarce. Rather this study focuses on the policy-formulation process and how and why Soviet siting policies with respect to nuclear power facilities have changed over time.

I have drawn largely from the conceptual work of Western geographers and location theorists who have written on the siting of hazardous or high-risk facilities such as nuclear power stations. This literature centers on decision-making and policy formulation, and specifically on who participates and how decisions are made concerning nuclear programs, projects, and locational choice. In examining the policy-formulation process in the Soviet context, this analysis is largely inspired by studies such as Donna Gold's on Soviet nuclear safety policy and Han-ku Chung's on regional energy development.[6] Studies of this nature rely upon careful interpretation of Soviet expert and public discussion as reported in professional and technical journals and in the popular press.

Specifically, I address the following questions:

(1) What societal groups participate through established institutions in the decision-making process? Has this changed over time?

(2) What societal groups are not represented institutionally? Have these groups been able to influence decision-making over time?

(3) How have the actions of institutions and other societal groups not institutionally represented influenced siting policy? How are these likely to affect the future of the industry?

(4) How has the Soviet experience with site decision-making and policy compared with Western experience?

Research Design and Sources

In tackling these questions, I had to depend on a wide range of sources. As mentioned earlier, this descriptive study relies heavily on the interpretation of official policy statements, as well as expert and public discussions and debates. For the convenience of others interested in the source materials, and because of my moderate familiarity with the Russian language, English language translations of Soviet source materials were used when possible.

Soviet sources can be divided into two categories: specialist literature and the popular press. Specialist or "expert" literature includes the energy journals *Soviet Atomic Energy* (an English language version of *Atomnaya Energiya*),* *Elektricheskiye stantsii* and *Teploenergetika*. Another source of expert discussion is the papers and reports submitted to the IAEA (International Atomic Energy Agency) concerning safety, siting, and reactor design issues. Other specialist literature includes books and published studies on the state of affairs and future plans in the energy sector printed

---

* *Atomnaya Energiya* was used in a few cases where articles were not published in the English language version *Soviet Atomic Energy*.

through the Soviet publishing houses *Nauka, Atomizdat,* and *Energoatomizdat.*

I relied on the popular press as a source to evaluate official policy positions of the national political leadership as well as local government and public attitudes and actions. Here I used primarily English language sources such as *Current Digest of the Soviet Press* (hereafter, referred to as *CDSP*), *Foreign Broadcast Information Service* (hereafter, *FBIS*), *Joint Publications Research Service* (hereafter, *JPRS*), and various press analyses and summaries from *Radio Free Europe/Radio Liberty* (hereafter, *RFE/RL*).

Western nuclear industry journals such as *Nuclear Engineering International, Nucleonics Week,* and others were also used, particularly as sources for opinions and attitudes of Soviet nuclear industry officials who at times appear to be more frank and forthright in their opinions and assessments with Western journalists than with the Soviet popular press.

Organization of the Book

A conceptual framework for the siting issue, drawing heavily on Western literature on Western experience, is presented in Chapter I. Chapter II describes the energy resource endowments of the Soviet Union and the relevant developments in the Soviet energy sector and provides a background for understanding the energy strategies adopted by national policy-makers and energy officials. The formal decision-making processes and participants during the early years of the nuclear power program until 1986 are identified and described in Chapter III. Soviet siting policies and the related issues of scale and safety technology from the early years of the nuclear power industry up to 1986 are described in Chapter IV. Changes in the formal decision-making participants and institutions, as well as siting policies in the nuclear industry, in the aftermath of the Chernobyl' accident, are described in Chapter V. The growth of public opposition and its impact on the siting issue as well as the spatial development of the nuclear industry in the Soviet Union between 1986 and 1990 are described in Chapter VI. Post-Soviet developments in nuclear power are briefly discussed in Chapter VII; and the Soviet experience in the nuclear facility siting issue is summarized and compared with the experience of several Western countries. For the reader's convenience, a glossary of acronyms and abbreviations is provided in the back of this volume. Additionally, appendices have been included providing data on individual nuclear stations and reactors that operated during the

Soviet period; types of public opposition to project sites during the period 1987–1991; a brief station by station historical description of both public opposition to and status of individual project sites; and a list of operational and planned stations and reactors in the newly independent states of the former Soviet Union.

Note on Transliteration and Place Names

Transliteration from Cyrillic to Latin characters is a confusing process and as a result there are several widely used systems. In this study I use the transliteration system of the United States Board on Geographic Names. For reference and comparison to other heavily used transliteration systems such as the United States Library of Congress and the British Standards Institution system refer to the 119th issue of the *Bulletin* published by the Library of Congress or any issue of *Soviet Geography*.[7] Throughout the text, wherever place names appear, the Board of Geographic Names system is used, except when other sources are cited directly and those sources use a different system. Certain place names, however, such as the "Crimea" the "Ukrainian SSR" or "Moscow" retain the commonly accepted English spelling.

Recent political events have contributed additional confusion to the use of place names. Unless otherwise indicated, original Soviet place names are used, although I have tried to point out name changes by including the currently used place name in parentheses. I feel compelled to do this for the convenience of the reader—many sources refer to the contemporary Soviet place name and these city changes occurred over a broad period between 1988 and 1991. Moreover, while it might be "politically correct" to use post-Soviet place names, the current state of political affairs in the independent states of the former USSR suggests that currently accepted place names might be transitory.

Notes

1. "Fewer Plants but More Capacity in 1990," *Nuclear Engineering International* (April 1991), p. 4.

2. Ibid.

3.   H. Kunreuther et al., *Risk Analysis and Decision Processes: The Siting of Liquefied Energy Facilities in Four Countries* (New York, NY: Springer-Verlag, 1983); Dorothy Nelkin and Micheal Pollak, "Consensus and Conflict Resolution: The Politics of Assessing Risk," in M. Dierkes et al. (eds.), *Technological Risk: Its Perception and Handling in the European Community* (Boston, MA: Oetgeschlager, Gunn, and Hain, 1979); Dorothy Nelkin and Micheal Pollak, *The Atom Besieged: Extraparliamentary Dissent in France and Germany* (Cambridge, MA: MIT Press, 1981); K. David Pijawka, "The Pattern of Public Response to Nuclear Facilities: An Analysis of the Diablo Canyon Nuclear Generating Station," in M. J. Pasqualetti and K. D. Pijawka (eds.), *Nuclear Power: Assessing and Managing a Hazardous Technology* (Boulder, CO: Westview Press, 1984); Herbert Kitschelt, "Political Opportunity Structures and Political Protest: Anti-Nuclear Protest in Four Democracies," *British Journal of Political Science*, Vol. 16, No. 1 (January 1986), pp. 57–85.

4.   Donna Gold, *Agenda Setting in Soviet Domestic Politics: The Case of Nuclear Safety Policy*, paper presented to the Annual Conference of the American Association for the Advancement of Slavic Studies, Honolulu, November 18–21, 1988; Judith Thorton, "Soviet Electric Power after Chernobyl': Economic Consequences and Options," *Soviet Economy*, Vol. 2, No. 2 (April–June 1986), pp. 131–179; William Kelly et al.,"Nuclear Electrification," in *Energy Research and Development in the USSR: Preparations for the Twenty-First Century* (Durham, NC: Duke University Press, 1986), pp. 48–82; William Kelly et al., "The Economics of Nuclear Power in the Soviet Union," *Soviet Studies*, Vol. 34, No. 1 (January 1982), pp. 43–68.

5.   Kunreuther et al.; Pijawka; D. J. Snowball and S. M. Macgill, "Coping with Risk: The Case of Gas Facilities in Scotland," *Environment and Planning C: Government and Policy*, Vol. 2, (1984), pp. 343–360; Steven L. Del Sesto, *Science, Politics and Controversy: Civilian Nuclear Power in the United States, 1946–1974* (Boulder, CO: Westview Press, 1979); Stan Openshaw, *Nuclear Power: Siting and Safety* (London: Routledge and Kegan Paul, 1986).

6.   Gold; Han-Ku Chung, *Interest Representation in Soviet Policymaking: A Case Study of a West Siberian Energy Coalition* (Boulder, CO: Westview Press, 1987).

7. Library of Congress, *The Bulletin*, No. 119 (Fall 1976), p. 63 (Washington D.C.: Library of Congress Cataloging Service); also see *Soviet Geography*, back of title page, various issues.

Chapter I

Conceptualizing the
Nuclear Facility Siting Issue

The object of this chapter is to examine the issue of siting nuclear power facilities as a location problem and to review conceptual issues brought up in previous research. The siting of a nuclear power-generating facility is a decision process in which different societal actors, sensitive to a wide range of location factors and criteria, interact in an institutional setting to resolve the siting decision. This chapter provides a conceptual framework for the siting issue, identifying the societal participants and processes involved in site decision-making. Such a framework is drawn from the research of several individual case studies as well as a few systematic comparative studies that together provide a generalized understanding of the factors and issues at work. It is hoped that through this framework differences and similarities can be identified between Western and Soviet experiences in the siting issue.

This chapter is organized in five sections. First, the unique characteristics and related issues of siting nuclear power facilities are identified and discussed. Second, from the literature on nuclear and other high-risk energy technologies with similar location characteristics, the various actors or decision-making participants and their respective decision criteria are identified. Third, the role and impact of different institutional settings or "environments" in which actors interact are discussed. This is followed by a discussion of the trade-offs involved in the siting decision and a brief overview of the geographic and temporal patterns of development for high-risk technologies such as nuclear power and how they might be influenced by changing perceptions and the participation of different societal actors.

Nature of the Siting Problem

The problem of locating nuclear power-generating facilities possesses some unique characteristics that have been described in a number of studies.[1] Of these, however, few have attempted to specify the nuclear power facility location problem as it relates to the general body of location theory.[2] Richetto, examining nuclear facility siting

9

in the United States, points out that the nuclear facility location problem straddles conventional conceptualizations of facility siting problems.[3] According to Richetto, the siting of a nuclear power facility shares characteristics of both private and public location problems. It resembles a public sector location problem in that the utility, the organizational entity responsible for initial site selection, produces an ordinary service (electricity),* which it is obliged to produce, while seeking to maximize the returns on its investment through revenue–cost optimization.[4] This characteristic of a utility or power ministry as a revenue–cost optimizing firm is not unique to capitalist countries.[5] Furthermore, because the utility or power ministry is given a fixed service region to supply, it acts as a spatial monopolist. Thus a utility's location decisions are made quite independently of other utilities' location decisions.**[6] Additionally, in selecting sites for large-scale power facilities, locations are not selected on a predetermined network design that optimizes transmission distances from producing to demand nodes. The nature of the technology used in power generation places significant constraints on the choice of locations. Thus, the search procedure for nuclear power-generating facilities is structured around the selection of appropriate locations within the specified demand area. (Indeed, this is typical of all large-scale electricity sources, such as thermal and hydroelectric plants.) Site-specific criteria include stable geologic conditions, adequate amounts of coolant water, adequate infrastructure, distance from heavily populated areas, etc.[7]

Risk presented to individuals and society through the use of nuclear power-generating technology adds additional complexity to the site selection problem. Nuclear fission technology is considered a high-risk technology, characterized by very low probabilities of potential accidents but with very damaging or lethal consequences resulting from such accidents.***[8] The level of risk associated with nuclear power can vary enormously depending on the type of fission technology used (e.g., pressurized-water, fast-breeder), as well as the attendant engineered safety systems. The location of the facility also influences the level of risk. Location factors influencing risk include the surrounding physical geography and the extent and type of

---

* Other examples of ordinary services include water, gas, and postal service.

** Utilities do at times share costs and information for joint projects.

*** A high-risk technology can also be characterized by high probabilities for potential accidents and relatively insignificant consequences resulting from such accidents.

human activity (air transport corridors, hazardous chemical processing facilities, etc.).[9]

Thus, site selection must consider factors that both increase and reduce risk. Additionally, because of the mechanics of radioactive contaminant dispersion in nuclear accidents, people living near nuclear generating facilities (i.e., less than 30 to 60 kilometers) are exposed to significantly higher individual risk than those farther away.[10] In addition to causing elevated health risks, nuclear plants can negatively affect land values, the aquatic environment, and the aesthetic beauty of the surrounding countryside. Thus, with the use of nuclear power-generating technology, there occurs a spatial variation of the benefits and risks and/or costs of the technology occurs. In this respect, nuclear generating facilities can be characterized as "noxious" facilities in which the residents in close proximity to the plant receive a disproportionate share of the negative effects of the facility (*real, potential,* and/or *perceived*).[11]

It can be seen, then, that in the case of site selection for nuclear power facilities, economic and noneconomic considerations of both a quantifiable and nonquantifiable nature come into play in the decision.[12] Included among the economic considerations are land acquisition, construction and infrastructure development costs, labor availability, wages, etc.[13] These types of factors lend themselves rather easily to quantification. Noneconomic factors are, however, generally less quantifiable. Some factors such as geologic stability, and the thermal effects of coolant discharge on aquatic life are, to a degree, measurable. On the other hand, factors, particularly "noxious" factors such as permissible levels of radiation and their effects on human, animal, and plant life, the aesthetic effects of cooling towers, the safety and reliability of fission technology, and the probability of hazardous releases of radioactivity, defy widely accepted, rigorous quantification, at least at current levels of human knowledge (although substantial headway has been made in recent years in reliability and risk assessment methodologies).[14] Nevertheless, these noneconomic or semiquantifiable factors can play an important role in the location decision; these factors are especially characteristic of problems in siting nuclear generating facilities. The extent to which they are recognized as important varies among different decision-making participants' perceptions and interests, and their importance in the final decision is largely determined by the institutional setting in which the decision takes place.

Decision-Making Participants and Their Criteria

Because many individuals and institutions view the siting of a facility as having a potentially significant impact on them, they attempt to influence the decision to conform to what they perceive as a favorable outcome. More and more studies in recent years have examined the decision-making processes and participants in the siting of high-risk energy facilities with "noxious" characteristics. These studies have primarily focused on site decisions with nuclear and liquefied natural gas (LNG) technologies and have contributed significantly to an understanding of who participates in the location decision.[15] Kunreuther et al. in a comparative international study of the siting process of liquefied natural gas facilities identified several actors that were common in each country examined (United States, the United Kingdom, the Federal Republic of Germany, and the Netherlands). This study concluded that there are four principal participants in the siting decision for LNG facilities: *the applicant* (firms, utilities, power ministeries, or developers who construct and operate the facility); *local residents* (residents living within a short distance from the proposed site); *government agencies* (government institutions with specified responsibilities at the national, state or municipal levels); *public interest groups* (politically active organizations representing the interests and preferences of certain sections of the public).[16] Studies by Snowball and Magill, Pijawka, Cook, Price, and others confirm the crucial role that utilities, local government bodies, local residents, and national public interest or "pressure" groups play in the siting decision.[17] Other studies, for example Revelle's, have identified *national policy-makers* as an important and distinct group of actors.[18]

Each of the actors participating in the decision-making process faces what Wolpert describes as a "decision environment," which is to say that each actor or decision-maker is sensitive to a different set of location criteria on which to base decisions.[19] In the following discussion, each of these actors is described in greater detail and their respective location decision criteria are discussed.

National policy-makers: This group includes national level planners and policy-makers. Typically, national policy-makers are concerned with regional development issues, national energy dependency, environmental quality, domestic political opinion, and in some cases even international political prestige. The decision-making criteria of national policy-makers, while appearing to be amorphous, are focused on national security, economic and political stability. Research suggests that national policy-makers typically are not involved in individual siting decisions, instead they initiate policy that profoundly affects the decisions of other actors. National policy-makers primarily affect the decision-making process by attempting to influence the power utilities or ministries through "objective function" methods (i.e., tax incentives, subsidies, user surcharges, effluent fees).[20] These methods are essentially financial incentives intended to influence the final outcome of the economic analysis of the utility or ministry. National policy-makers also influence siting decisions by means of "administrative" methods. Administrative methods include laws, for example national siting legislation, or actual "commands" to the utilities or ministries to base additional capacity expansion upon a type of technology. These methods of intervention by national policy-makers have frequently been deemed necessary as the uncertainty and lack of experience with nuclear technology have been powerful disincentives for the adoption of this technology.

Utilities or power ministries: As pointed out earlier, the utility or power ministry is the actor that is responsible for initial site selection and plays a key and central role in the decision-making process. Typically, each utility or power ministry supplies all electrical demands within a well-defined service region. Thus a power utility or ministry is a regulated monopoly entrusted with the responsibility for providing electrical service to the public. At the same time, utilities seek to optimize their returns on the investment associated with their siting and technology decisions.[21] Consequently, for power utilities or ministries, siting criteria revolve around engineering parameters (adequate cooling water, topography, geology, and available land) that must be met and around economic factors (costs of land acquisition, construction, and fuel transport, as well as costs associated with additional transmission lines, labor availability, projected demand, and local tax structures), which are assessed through some form of revenue–cost optimization.[22] Additionally, the utility or ministry, depending upon the existing institutions and procedures, must consider the extent and seriousness of potential public opposition, which could result in very costly delays in construction. Utilities' criteria are thus economic and technical and are therefore more susceptible to some form of quantitative analysis.

Local public: Considerable research exists on the composition of opposition groups to nuclear and other hazardous technologies. Kunreuther et al., Price, and others make an important distinction between local residents and nationally focused "public" interest groups in both the role they play and their interests.[23] Research on the U.S. experience suggests that while a majority of the American public supported nuclear power in the abstract, even in the wake of the Three Mile Island accident in 1979, a large segment of the same public was strongly opposed to the location of nuclear plants near their homes.[24] Such sentiment, also known as NIMBYism from the acronym NIMBY (Not In My Back Yard), exists due to the "noxious" nature of the facility. Some residents may be concerned with decreases in their property values as a result of the new power plant or the negative aesthetic effects of the facility on the surrounding countryside. Individuals also may be disturbed by the possible health risks associated with the facility's close location, and especially the consequence of catastrophic high-level radiation releases as a result of a reactor accident.

Other research, however, illustrates that NIMBYism is not a universal reaction on the part of local communities. Communities actually may be favorably inclined to the new facility because of such factors as increased local tax revenues, employment, and local business activity.[25] The perception and importance of these factors can vary enormously among individuals and communities. Several commentators have attributed variations in the perceptions and attitudes among local residents to levels of education, local employment conditions and the state of economic development, as well as public trust in the competence of technical experts and government and utility officials.[26]

Public interest groups: This category of actors consists of regional or national citizens' or environmentalist groups. For instance, examples from the U.S. experience would include the Sierra Club, the Union of Concerned Scientists, and the National Resources Defense Council, to name a few.[27] These groups differ from local residents in that they do not necessarily share the local interests, although their interests frequently do coincide. Instead, public interest groups are typically regionally or nationally based organizations concerned with a single aspect of the siting problem, for example, environmental impacts or the dangers of nuclear technology.[28] While such groups' interests may coincide with those of local residents, as Pijawka and Kunreuther et al. have shown, there may also be a divergence of interests between these two groups.[29]

Local government agencies: This category of actors consists of regional, state, and municipal government bodies. This group differs from national policy-makers in that their interests are more narrowly defined within a smaller geographic area. Typically this group of actors becomes involved in the siting decision when they attempt to influence decision-making by imposing restrictions or benefits in legislative form on the power utilities or ministries. Past research has shown that local government agencies are sensitive to a wide range of considerations, including regional electricity availability, increased employment opportunities, enhanced tax bases, and constituent attitudes toward the nuclear facility.[30]

The Institutional Setting and the Decision-Making Process

The relative importance that each of these groups of actors plays in the decision-making process is largely determined by the institutional environment, consisting of the established institutions and procedures through which they interact. Several studies on the decision processes involved in the siting of nuclear power and other energy facilities with noxious characteristics indicate that the institutional environments vary considerably among Western countries. According to Dooley et al., the relative power of the various participating actors in the decision process is determined by the established institutions and procedures found in each country.[31] The research of Nelkin and Pollak, Kitschelt, and others further illustrate the significance of institutional structures in nuclear power decision-making.[32] Nelkin and Pollak, in their analysis of decision-making with high-risk technologies in North America and Europe, describe various institutions and procedures that have developed in different countries as those technologies have become increasingly controversial.[33] According to Nelkin and Pollak, institutions and procedures involved in decision-making with high-risk technology fulfill two important roles in society: (1) relaying "advice" to decision-makers, including levels of risk acceptable to the public, nonmonetary values on land use, environmental disruption, etc.; and (2) relaying information to the public, such as estimated risks of the new technology, the economic and employment benefits to be gained, and the costs and risks of alternative technologies.

Thus there is an attempt to bridge the quantitative technical and economic considerations with non- or semiquantifiable value questions that affect different actors through institutions and the decision-making processes and procedures. In their analysis, Nelkin and Pollak present two conceptual models that describe the structure of a society's institutional environment for high-risk technology decision-making. In the "elitist" model, conflicts surrounding controversial decisions on high-risk technologies are resolved by select groups of experts. Under such a model, consensus is reached between experts representing national policy-makers, utilities, and sometimes local government officials with little or no public or "non-expert" participation in the decision-making process. Several studies indicate that nuclear facility siting processes in the Uuited Kingdom, Canada, and France resemble this model.[34] An alternative model put forth by Nelkin and Pollak is the "participatory" model. Within the participatory framework, different elements of the public become involved in the decision-making process. Typically, public participation in this model takes place through institutions and procedures, such as public inquiries, licensing hearings, and advisory councils. In some cases, elitist and participatory institutions and procedures may coexist while performing different functions. For example, participatory procedures and institutions may serve an advisory function, while elitist procedures and institutions are established to disseminate information to the public.[35]

Kitschelt's comparative research on public opposition's role and impact on the nuclear programs in France, Sweden, the United States, and the Federal Republic of Germany provides a further conceptual explanation of the relationship between institutional environments and the influence of opposition groups in decision-making. Although Kitschelt's research does not focus on the siting issue specifically, his analysis nevertheless is instructive in providing an understanding of the different institutional environments involved with nuclear power in these countries. Kitschelt uses the term "political opportunity structure" to define the key set of variables that determine the actions and effects of opposition movements on decision-making and policy formulation. "Political opportunity structure" includes the institutional environment in which opposition groups operate; the character of this structure influences the choice of protest strategies and the impact of social movements on decisions, policies, or programs.[36] In Kitschelt's analysis, opponents can affect decisions via two avenues—the input or participatory side and the output or policy implementation side. The relative openness of political regimes to public input and the strength of the political regime in implementing policies are the key characteristics that define political opportunity structures.* The result of relatively closed input structures is a

---

* Kitschelt states that the relative "openness" of political regimes is determined by such factors as the number of political parties or factions that can effectively articulate demands, and the ability of legislatures to initiate and control policies independently of the executive and the procecural environment that facilitates the formation of policy coalitions. The strength of a political system in implementing

tendency toward "confrontational" strategies—disruptive activities that are designed for political effect and conducted outside established policy channels (public demonstrations, acts of civil disobedience, such as occupations of nuclear plant sites and access roads). Conversely, open structures generate "assimilative" strategies—opponents of projects attempt to work through established institutions (legislative lobbying, petitioning government bodies, referendum campaigns, etc.).[37] According to Kitschelt, the United States and Sweden closely resemble open input structures.

Equally important, the public can influence decisions at the policy implementation stage. In countries with weak implementation capacities, such as the Federal Republic of Germany or the United States, a multitude of competing regulatory agencies, an open licensing process, and an independent judiciary have helped delay many nuclear projects, thus increasing costs and negatively affecting the growth of the industry.[38]

Wood's research on the U.S. regulatory experience further illustrates the relationship between institutional structures and decision-making. In the U.S. system, licensing according to Wood, provides an open, adversarial atmosphere with few decisions "set in stone" and with design changes always a possibility. Wood points out that, in contrast, in most European countries technical licensing is conducted out of public view. With the exception of the Federal Republic of Germany, opposition to nuclear projects and policies is voiced in a political forum rather than through specific licensing actions.[39] Cook and others have also noted that in the United States and the Federal Republic of Germany, local interests have had much more influence in affecting specific facility decisions, as well as the as overall course of the nation's nuclear industry primarily by using the licensing process and the court system.[40]

Even within a country, there may be variations in the institutional environment. This occurs in federal systems such as the United States where local governments (i.e., state or county) have to a limited degree imposed their own institutions, procedures, and regulations. Indeed, in the United States some states have allowed greater public participation, such as mandatory licensing hearings for proposed facilities; while local regulations on site acceptance vary enormously from state to state.[41]

---

policies is dependent on such factors as the centralization of the state apparatus and the degree of government influence over market participants, as well as the relative independence and authority of the judiciary in resolving political conflict. (Kitschelt, pp. 62–64)

Moreover, active constituencies have been able to induce local legislative and executive officials to enact legislation or intervene in the licensing procedure to change a utility's decision to build a nuclear power station. In the United States, for example, more restrictive nuclear legislation was passed in six states in the early 1980s as a result of state referenda.[42] Local governments have been known to utilize legislative procedural loopholes to frustrate a utility's plans, even after the siting proposals have passed successfully through established institutional processes.[43]

Trade-off and Compromise in the Siting Decision

Location theorists recognize that location decisions result from compromises and trade-offs between interested actors and participants involved in the decision-making process.[44] This is especially true in the case of "noxious" facilities, such as highways, airports, or power stations. Mumphrey et al. suggest that typically while at an aggregate regional level, cost-benefit analysis may justify the location of a facility at some site within a region, alternative communities may compete not to have it located in their vicinity. Locating such facilities implies intercommunity transfer payments whether these be actual or perceived consequences.[45] Mumphrey et al. and others have pointed out that in locating controversial noxious facilities, trade-offs or compromises usually take the form of location changes or sidepayments to the affected community or set of communities.[46] For a high-risk technology such as nuclear power, sidepayments are frequently perceived as an inappropriate form of compensation. Instead in the case of nuclear power, compromises between decision-making participants can involve location, scale, and technology.

The trade-offs involved in the siting decision take into account four broad concerns: (1) different sites expose different people to hazards; (2) different sites expose different numbers of people to different hazards; (3) different sites involve different types and levels of hazard; and (4) different sites have different resource endowments and thus offer different cost advantages.[47] The first concern refers to the fact that different communities often vary in their attitude toward a nuclear power facility and that these variances in opposition can play an important role in the final decision. The second concern relates to the fact that sites in close proximity to population centers increase the overall risk of the technology. Because close proximity to population concentrations increases the number of individuals exposed to the risk of contamination, the potential for opposition is greater. From society's point of view, the ideal site would expose the fewest number of individuals to risk. Different sites also present differences in the types and levels of hazard, for example, different population densities. Individual sites also possess natural or anthropogenic hazards, which add to the risk of a potential accident. Natural hazards include seismic frequency and magnitude, geologic stability, and local meteorological conditions. Anthropogenic hazards include such things as the probability of airplane crashes, chemical explosions, related to the facility's proximity to air-transport corridors, roads, and industrial activities.[48] A potential site must also include resources that do not make the facility's construction and operation prohibitively expensive. Certain resources are absolutely necessary for a facility's operation; these include an adequate supply of cooling water and stable, level land. Adequately developed infrastructure, available labor, and spatial proximity to demand centers are other site resources or characteristics that greatly reduce the cost of facility construction and operation.

Another set of trade-offs includes the "safetyness" of the reactor technology or, more specifically, the degree of safety provided by the reactor and facility designs. The degree of safety that a design provides is a product of three components: the intrinsic safety characteristics of the reactor design, and the accompanying passive and active safety systems. Improvements in each of these safety components entails additional costs.

The distinctions among intrinsic, passive and active safety are important because the reliability and safety assuredness of the first two are much higher than the third. Intrinsic safety is achieved by engineering the physical characteristics of the reactor so that the possibility of an accident and subsequent damage to the reactor as a result of an accident are prevented as much as possible. Passive safety features or systems are designed to counteract accident conditions without active human intervention. This is achieved through such features as pressure suppression pools and containment vessels (to hold radioactivity released in an accident). Active safety systems are engineered safety systems that monitor and convey information to human operators and interrupt reactor processes upon human initiation. Active safety systems are considered a less reliable component of reactor safety because they rely heavily on human competence, as well as mechanical and electrical components, which are subject to failure. Although reactors with good intrinsic safety characteristics and several redundant active and passive features may greatly increase the margin of safety, they frequently reduce operating performance and involve greater capital expenditures and higher operating costs.

Reactor and facility scale represent another set of trade-offs that can be considered in the location decision. Reactor and facility scale affect cost, risk, and the physical input requirements of the facility. Both the scale of the reactor and the scale of the facility present a number of options (many small reactors or a few large reactors at a specific site). With respect to reactor scale, large units offer greater economies of scale with lower capital costs per unit for installed capacity and output. Nevertheless, some advantages are associated with small reactors. Smaller units have a smaller inventory of radionuclides available for potential release. Smaller reactor size can in some cases also reduce the possibility of a meltdown, depending on the reactor design used.[49]

Different scales of plant or facility size offer a variety of advantages and disadvantages with respect to economy and safety. Larger plants economize capital expenditures associated with site preparation and infrastructure development, as long as plant capacity does not exceed consumer demand and available water resources. Alternatively, smaller facilities offer an advantage in that they present less of a demand on local resources, thus reducing their ecological impact and extending the range of possible sites. Facility size also conjures up the issue of risk and equity, given that a larger plant (assuming a larger number of reactors) increases risk in a specific area compared with a smaller plant. Such a relative concentration of risk to a given population may be perceived as socially unfair.[50]

Decision-Making and the Geographic and Temporal Patterns of Nuclear Industry Development

The perception of the participants in the decision-making process as well as in the institutional environment may indeed have a significant effect on the nuclear industry's pattern of development. Semple and Richetto hypothesize that any innovative private but publicly regulated industry that uses a complex experimental technology with perceived negative externalities will, by its very nature, be subject to an identifiable number of distinct locational phases.[51] Semple and Richetto call these phases experimental, rapid expansion, retrenchment, and controlled or regulated expansion. During the initial or experimental phase, a few small pilot facilities are located near traditional facilities with accepted proven technology. These pilot facilities are used primarily for demonstration and experience. The second phase encompasses a period of rapid expansion. During this time, there is a rapid geographic dispersal of the technology and transition to utilization at scales comparable to traditional technology. This phase is characterized by a "bandwagon" effect in which the rapid adoption of this technology results from the decision-makers' desire for security against uninterrupted supply, the establishment of a secure manufacturing and technical base for future expansion, and from the enthusiasm always present to acquire new and prestigious technology. The third phase is one of retrenchment of the technology in question. A crucial point is reached at this time because the rapid expansion of the technology brings attention to itself. During this phase, public concern over the expansion without proper assessment and critical examination of the potential impacts leads to a broadening of the scope and number of participants involved in the decision-making process and consequently restrictions are placed on innovation. These restrictions might include measures such as new environmental safeguards, increased public regulation, and private user rationalization. If the technology survives public reassessment then the fourth phase follows. During this phase, expansion continues but at a controlled and balanced rate.[52]

Although in the abstract Semple and Richetto's phases might hold up to reality, they still fais to explain the variations between countries. Indeed such an orderly pattern is not necessarily borne out over the years. Among the Western countries, the experience of France and Japan represents fairly continuous growth. Nevertheless, these patterns have been borne out elsewhere, such as in the United States, the United Kingdom, the Federal Republic of Germany, and Sweden, although for more complex reasons than the explanation Semple and Richetto provide. Factors unrelated to heightened public concern such as unforeseen costs (although certainly in the United States and the Federal Republic of Germany, opposition-induced delays have exacerbated costs), unfavorable capital markets, and reductions in the demand for domestic electricity have contributed to the retarded growth of the nuclear industry in many countries.

Jasper and other social scientists have even concluded that nuclear policies in the United States, France, and Sweden changed not because of internal or external factors such as public opposition, rising construction costs, or changing energy prices, but because the preferences of the policy-makers changed. In Jasper's words: "public opinion, plant costs, and safety, alternative energy sources, the strength of the antinuclear movement, legislation and regulation were all transformed to fit chosen policy paths."[53] Reality lies somewhere in between Semple and Richetto's conception of a universally recognizable path of industry growth with the gradual imposition of restraints on the technology and Jasper's that the path of development for the industry is primarily determined by national policy-makers. Indeed much of the research cited above indicates that in the struggle between an expanding industry with a plethora of supporters and interests and those concerned with the externalities and effects of that industry, and the structure of the institutional environment do, to a large extent, determine the character of the technology and its use.

Notes

1. Ralph Keeney, *Siting Energy Facilities* (New York, NY: Academic Press, 1982); A. E. Green, *High Risk Safety Technology* (New York, NY: John Wiley & Sons, 1982); Jeffery P. Richetto, "Locating Nuclear Electric Energy Facilities: Structural Relationships and the Environment," in Pasqualetti and Pijawka, pp. 103–119; John W. Winter, *Power Plant Siting* (New York, NY: Van Nostrand Reinhold Co., 1978).

2. Richetto, p. 103.

3. Ibid., p. 104.

4. Ibid., pp. 106–107; Charles Revelle et al., "An Analysis of Private and Public Sector Location Models," *Management Science*, Vol. 16, No. 11 (July 1970), pp. 692–693.

5. Katherine, Young (ed.), *Decision Making in the Soviet Energy Industry* (Falls Church, VA: Delphic Associates, 1986); David Dyker, *The Process of Investment in the Soviet Union* (Cambridge: Cambridge University Press, 1983).

6. Richetto, pp. 106–107.

7. Ibid.; Revelle et al., pp. 693–695.

8. For a comprehensive discussion on the concept of industrial and technological risk, see William McCormick, *Reliability and Risk Analysis: Methods and Nuclear Power Applications* (New York, NY: Academic Press, 1981), pp. 231–235; and W. Vinck,

"Practices and Rules for Nuclear Power Stations: The Role of the Risk Concept in Assessing Acceptability," in Dierkes et al., pp. 111–112.

9. Keeney, pp. 80–84; B. K. Grimes, "External Hazards As They Affect Nuclear Power Plant Siting," in *Siting Nuclear Facilities* (Vienna: IAEA, 1975), pp. 140–144.

10. Populations residing close (less than 30–60 kilometers) to nuclear power facilities are exposed to a higher possibility of contamination in the event of an accident than if plant were farther away. This is due to the fact that the most effective and the most likely means by which individuals become exposed to radiation is in the form of atmospheric releases. Distance from the source of release helps protect potential victims from the possibility of exposure in a variety of ways. First, through the processes of dry and wet desposition (fallout), radioactive decay (some radionuclidies have half-lives of a few hours), and dispersion, the concentration of radionuclides and their absolute number decreases with distance. Moreover, typically the greater the distance from the point of release, the more time is available to take protective measures such as evacuation or staying indoors. T. F. John, "Environmental Pathways of Radioactivity to Man," in *Nuclear Power Technology, Vol. 3: Radiation* (Oxford: Clarendon Press, 1983), pp. 155–216; Peter Mounfield, *World Nuclear Power* (New York, NY: Routledge, 1991), pp. 261–289.

11. A "noxious" facility is defined as a facility that has significant differences in the incidence of costs and benefits. Such facilities yield harmful neighborhood effects to some population subgroups. A. Mumphrey et al., *A Decision Model for Locating Controversial Facilities*, Discussion Paper No. 11 (Philadelphia: Department of Geography, University of Pennsylvannia), p. 2. Also see Bryan Massam, *Spatial Search: Applications to Planning Problems in the Public Sector* (Oxford: Pergammon Press, 1980).

12. Richetto, pp. 105–107.

13. Ibid.

14. For more detailed discussion in the area of risk with respect to methodology and policy implications, see McCormick, *Reliability*

and Risk Analysis; Roger E. Kasperson and Jeanne X. Kasperson, *The Impacts of Large-Scale Risk Assessment in Five Countries* (Boston, MA: Allen & Unwin, 1987).

15. These studies include multi country comparative studies: Kunreuther et al.; Nelkin and Pollak, "Consensus and Conflict Resolution." Or single-country studies in case study format: Pijawka, "The Pattern of Public Response"; Snowball and Macgill, pp. 343–360; Steven Ebbin and Raphael Kasper, *Citizen Groups and the Nuclear Power Controversey: Uses of Scietific and Technological Information* (Cambridge, MA: MIT Press, 1974); Constance Ewing Cook, *Nuclear Power and Legal Advocacy* (Lexington, MA: Lexington Books, 1980).

16. Kunreuther et al., p. 10

17. Snowball and Macgill, pp. 343–360; Pijawka, "The Pattern of Public Response," pp. 215–220; Cook, pp. 9–24, 49–60; Jerome Price, *The Antinuclear Movement* (Boston, MA: Twayne Publishers, 1990), pp. 27–35.

18. Revelle et al., pp. 692–695; Massam, pp. 56–60.

19. Julian Wolpert, "Departures from the Usual Environment in Location Analysis," *Annals of the American Association of Geographers*, Vol. 60, No. 2 (1970), pp. 220–221.

20. Revelle et al., pp. 692–695; Massam, pp. 56–60.

21. Richetto, pp. 106–107.

22. Ibid.; Mounfield, pp. 251–260.

23. Cook, pp. 9–15, 50–56; Price, pp. 28–30; Ebbin and Kasper, pp. 187–223.

24. Barbara Farhar-Pilgrim and William Freudenburg, "Nuclear Energy in Perspective: A Comparative Assessment of the Public View," in William Freudenburg and Eugene A. Rosa (eds.), *Public Reactions to Nuclear Power: Are There Critical Masses?* (Boulder, CO: Westview Press, 1984), pp. 183–189.

25. Richetto, pp.106–107; Pijawka, "The Pattern of Public Response," pp. 215–222.

26. K. David Pijawka, "Public Sector Effects and Social Impact Assessment of Nuclear Generating Facilities: Information for Community Mitigation Management," in Pasqualetti and Pijawka, pp. 181–187.

27. Price further categorizes national level public interest groups into scientific or technological antinuclear groups, moral activist groups, and direct action groups. Price, pp. 27–31, 37–101.

28. Kunreuther et al., pp. 10–11.

29. Pijawka, "Public Sector Effects," pp. 185–187; and Pijawka, "The Pattern of Public Response," pp. 217–219, 228–229.

30. Ibid.; Kunreuther et al., pp. 10–177; Nelkin and Pollak, "Consensus and Conflict Resolution," p. 74.

31. James E. Dooley et al., "The Management of Nuclear Risk in Five Countries: Political Cultures and Institutional Settings," in Kasperson and Kasperson, pp. 27–48.

32. Nelkin and Pollak, The Atom Besieged; Kitschelt; Cook.

33. Nelkin and Pollak, "Consensus and Conflict Resolution," pp. 65–75.

34. Ibid.; Dooley et al., pp. 46–48; John Fernie and Stan Openshaw, "Policy Making and Safety Issues in the Development of Nuclear Power in the United Kingdom," in Pasqualetti and Pijawka, pp. 68–69.

35. For example, France's Complaint Investigations Déclaration d' Utilité Publique consisting of concerned scientists, citizens, and public officials fulfills an advisory role in a participatory framework; whereas the Conseil d' Information sur l' Energie Electro-Nucleaire, a select group of national level officials, scientists, and ecologists, oversee and recommend what information should be publically released, thus representing an informative role in an elitist framework. Nelkin and Pollak, "Consensus and Conflict Resolution," pp. 66–69.

36. Kitschelt, p. 58.

37. Ibid., pp. 66–67.

38. Ibid., pp. 70–71, 79.

39. William Wood, *Nuclear Safety: Risks and Regulation* (Washington, D.C.: American Enterprise Institute for Policy Research, 1983), p. 31.

40. Cook, pp. 9–24, 49–60; Kitschelt, pp. 67–73; Nelkin and Pollak, *The Atom Besieged*, pp. 34–60.

41. Kunreuther et al., pp. 10–77; Winter, pp. 28–32; Cook, pp. 6, 39–40; Cook also attaches significance in the United States to the selection of one of the eleven regional Federal Courts of Appeals in a case when legal action is taken. She cites the Court of Appeals for the District of Columbia Circuit as a liberal court, more predisposed toward a favorable ruling in a case involving environmental issues. Cook, p. 45.

42. These states are California, Connecticut, Maine, Montana, Oregon, and Washington. Eugene Rosa and William Freudenburg, "Nuclear Power at the Crossroads," in Freudenburg and Rosa, pp. 24–25.

43. The Perry-1 and Davis-Besse nuclear stations in Ohio, the Seabrook plant in Massachusetts, and the Shoreham plant in New York are all examples where executive and legislative branches of state government have used *ad hoc* means to delay or obstruct nuclear projects. "Ohio Governor Pulls Support for Perry and Davis-Besse Emergency Plans," *Nucleonics Week*, Vol. 27, No. 34 (August 21, 1986), pp. 1–2; "Seabrook Startup Delays Have Cost $300 Million," *Nucleonics Week*, Vol. 27, No. 23 (May 29, 1986), p. 14.

44. R. Keith Semple and Jeffery P. Richetto, "The Location of Electric Energy Facilities: Conflict Coalition and Power," *Regional Science Perspectives*, Volume 9, No. 1 (1979), pp. 117–138.

45. Mumphrey et al., p.1.

46. Ibid., pp. 1–4.

47. This conceptualization is taken in part from Thomas Wilbanks, "Scale and the Acceptability of Nuclear Energy," in Pasqualetti and Pijawka, p. 9.

48. Grimes, pp. 140–144.

49. William Board, "Experts Say Reactor Design is 'Immune' to Disaster," *New York Times*, November 15, 1988, pp. 25–26.

50. See K. S. Shrader-Frechette, *Science Policy, Ethics and Economic Methodology: Some Problems of Technology, Assessment and Environmental-Impact Analysis* (Boston, MA: D. Reidel Publishing, 1985), pp. 220–238.

51. R. Keith Semple and Jeffery P. Richetto, "Locational Trends of an Experimental Public Facility: The Case of Nuclear Power Plants," *The Professional Geographer*, Vol. 28, No. 3 (August 1976), pp. 249–250.

52. Ibid.

53. James M. Jasper, *Nuclear Politics: Energy and the State in the United States, Sweden and France* (Princeton, NJ: Princeton University Press, 1990), p. 264.

Chapter II

Overview of the Soviet Energy Sector

This chapter provides an overview of the developments in the USSR's energy sector during the Soviet period so as to clarify the constraints and options that have faced planners and policy-makers in the course of developing that sector of the economy. The emergence of the nuclear power industry in the USSR was a consequence of long-term trends occurring within the energy sector. This chapter describes the conditions under which decisions were made concerning nuclear power development and siting policy. It also focuses on primary fuel reserves and production as well as on growth and consumption trends in two consumer areas of primary fuels in which nuclear power is reasonably seen as a substitute or alternative energy source—electricity and heat production.*

The Soviet Union has made several shifts in strategy in the course of developing its energy sector. This has primarily been due to increasing costs and production constraints among the different sources of conventional fuels, new resource finds, and rapid growth in demand. The geographic component of these problems strongly influenced the development of nuclear power.

Location and Extent of Soviet Energy Resources

The region encompassing the former Soviet Union is one of the most favorably endowed in the world with respect to energy resources, at least in aggregate terms. Its energy reserves are not only large but well balanced among primary fossil fuels, water resources suitable for hydroelectric development, and renewable fuels such as peat and wood. Although estimated energy reserves have changed, by the early 1970s, the USSR was estimated to rank

---

* Discussion on a third user area—transport—is omitted because it is not seen as a significant area of application and is not related to the siting issue. The Soviets have made use of nuclear energy for ship propulsion on some civilian ships, including icebreakers and merchantships. To an extent, nuclear ships do present a location issue because there has been public opposition in Japan and the USSR to servicing such vessels in local ports. However, these issues lie outside the realm of this study.

fourth in proven oil reserves, third in proven coal reserves, and first in proven natural gas reserves.[1] Since then the USSR's position actually improved; in 1990 the USSR was estimated to rank first in the world in natural gas reserves, a solid second in coal reserves and water resources with hydroelectric potential, and fourth in oil reserves.[2] Offsetting these advantages, the USSR was severely disadvantaged in the location of these resources. A marked spatial dichotomy existed between regional energy demand and energy supply. Over time, this spatial dichotomy became even more pronounced as demand grew and extraction activities depleted more accessible energy sources.

Historically, energy demand has been concentrated in the European USSR. Throughout the Soviet period, at least 75% of the Soviet population resided in the European USSR (including the Urals and Trans-Caucasian economic regions).[3] In addition, Soviet industry was located in the European USSR, especially around Moscow, and in the Donets, Dnepr, and Volga river basins. As of 1970 about 73% of gross industrial production was concentrated in the European USSR, 82% if the Urals are included in this figure.[4] By the early 1990s, these relative figures had not changed significantly.

The distribution of the former Soviet Union's major primary fuel reserves including coal, oil, natural gas, and gas condensate are shown on Maps 1 and 2. Soviet coal reserves are widely distributed throughout the country (Map 1). However, nearly 75% of the USSR's explored coal reserves are located east of the Urals.[5] Most of these eastern coals lie under the swamplike Siberian taiga or in Arctic permafrost far from demand centers. Only three large fields are well located with respect to European demand centers, these include the Moscow, Dnepr, and Donets basins. Of these only the Donets possesses high-quality hard coals of high calorific content. Among the Siberian deposits, the Kuznets and Lena basins are of considerable size and quality. Of these, however, only the Kuznets is reasonably accessible.

The vast majority of Soviet gas and oil reserves were concentrated in West Siberia, particularly in the supergiant Tyumen and Urengoy fields (Map 2). Other sizable oil and gas deposits include the Ukrainian gas fields, the North Caucasus oil and gas fields and the Volga–Urals oil fields. However, these smaller fields, while located closer to demand centers, have been depleted over time, lowering their importance as sources of supply.

The distribution of the USSR's water resources with hydroelectric potential are concentrated in Siberia and to a lesser extent Central Asia. Of an estimated total of 1,095 billion kilowatt

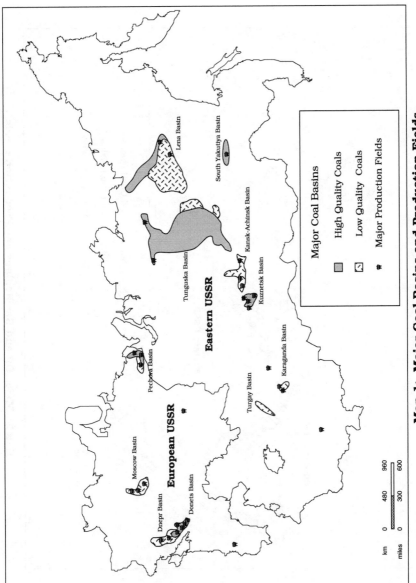

Map 1: Major Coal Basins and Production Fields

Major Coal Basins

High Quality Coals

Low Quality Coals

Major Production Fields

Lena Basin

South Yakutya Basin

Kansk-Achinsk Basin

Kuznetsk Basin

Tunguska Basin

Eastern USSR

Karaganda Basin

Turgay Basin

Pechora Basin

Moscow Basin

European USSR

Dnepr Basin

Donets Basin

km   0   480   960

miles 0   300   600

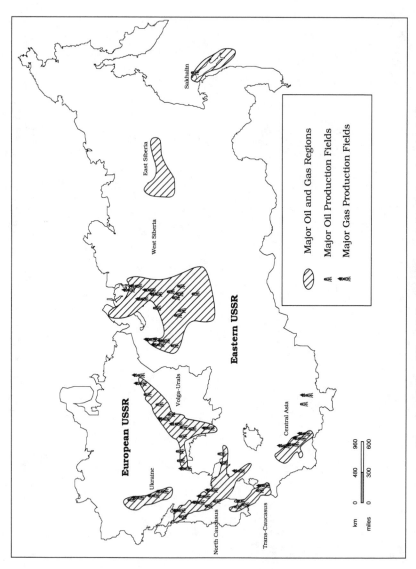

Map 2: Major Oil and Gas Regions and Production Fields

Sakhalin

East Siberia

West Siberia

Eastern USSR

Volga-Urals

Central Asia

European USSR

Ukraine

North Caucasus

Trans-Caucasus

Major Oil and Gas Regions

Major Oil Production Fields

Major Gas Production Fields

km   0   480   960

miles 0   300   600

hours (kWh) of annual electric power output that could be produced economically from the USSR's rivers, some 65.8% is in Siberia and the Soviet Far East and another 13.3% in Soviet Central Asia (not including Kazakhstan). Less than 20% of the USSR's potential hydroelectric capacity is found in the European USSR.[*] [6]

Development of Conventional Primary Fuels

Shortly after achieving power in 1917, the Communist Party placed energy development high on its economic agenda, where it remained throughout the Soviet period. Of particular importance to the party leadership was the development of fossil fuels to supply the growing demands of industry. As can be seen in Tables 1 and 2, Soviet energy production was based heavily on wood and coal during the early 1920s. However, as the USSR embarked on rapid industrialization during the late 1920s and 1930s, a shift to coal as the main source of primary energy occurred (Tables 1 and 2). Indeed, coal production expanded at an astonishing rate, more than quadrupling between the beginning of the 1st Five-Year Plan in 1928

Table 1: Soviet Primary Fuel Production, 1922–1990

Fuel	Output in million tons of standard fuel[a]						
	1922	1940	1960	1970	1980	1985	1990
Oil[b]	6.7	44.5	211.4	502.5	862.6	851.3	816.2
Gas	—	4.4	54.4	233.5	514.2	742.9	941.1
Coal	9.0	140.5	373.1	432.7	476.9	439.8	425.5
Peat	0.9	13.6	20.4	17.7	7.3	5.5	3.7
Shale	—	0.7	4.8	8.8	11.8	10.2	8.6
Wood	13.1	34.2	28.7	26.6	22.8	23.4	18.9
Total	29.7	237.9	692.8	1,221.8	1,895.6	2,073.1	2,214.0

— less than 0.1

[a] Based on standard fuel equivalents, 1 kg equals 7,000 kilocalories.

[b] Includes gas condensate.

Sources: 1922–1980 data from TsSU, *Narodnoye khozyaystvo 1922–1982*, p. 181; 1985 data from Goskomstat, *Promyshlennost' SSSR*, p. 141; 1990 data from Goskomstat, *Narodnoye khozyaystvo SSSR v1990*, p. 397.

---

[*] By 1990, annual output of hydroelectric power nationwide amounted to 233 billion kilowatt hours.

Table 2: Balance of Soviet Primary Energy Production, 1922–1985

Fuel	1922	Percent share[a] 1960	1970	1980	1985
Oil[b]	22.5	29.4	39.6	43.6	38.8
Gas	0.1	7.6	18.4	26.0	33.9
Coal	30.5	51.9	34.1	24.1	20.1
Peat	3.0	2.8	1.4	0.4	0.2
Shale	—	0.7	0.7	0.6	0.5
Wood	44.1	4.0	2.1	1.1	1.1
Electricity[c]	—	3.6	3.7	4.2	5.4
Total	100	100	100	100	100

— less than 0.1

[a] Based on standard fuel equivalents as reported from Soviet sources.

[b] Includes gas condensate.

[c] Primary electricity, includes electricity produced from nuclear fission and hydroelectricity.

Sources: Conventional primary fuel data from TsSU, *Narodnoye khozyaystvo, 1922–1982*, p. 181 and *Promyshlennost' SSSR*, p. 141; primary electricity from Goskomstat, *Material'no-tekhnicheskoye obespecheniye narodnogo khozyaystva, SSSR*, p. 64; and A. S. Pavlenko and A. M. Nekrasov (eds.), *Energetika v SSSR 1971–1975 godakh*, p. 36.

and the end of the 3rd Five-Year Plan in 1940, from 35.5 million metric tons.[7] The majority of this expansion in production occurred in European fields, particularly in the Donets Basin, which in 1940 accounted for 57% of Soviet coal production. The Soviet coal industry continued to maintain its role as the country's primary fuel source well into the 1960s. Growth rates for coal production began to decline noticeably during the 1960s and early 1970s. Despite the optimistic hopes of planners for revived growth during the 9th Five-Year Plan (1971–1975) and especially the 10th Five-Year Plan (1976–1980), the industry performed poorly, with output levels virtually stagnating during the 10th Five-Year Plan. This stagnation continued into the 11th Five-Year Plan (1981–1985) and 12th Five Year Plan (1986–1990).[8]

Soviet planners had hoped to increase the role of coal in the energy balance in the 1970s and 1980s in order to free up oil and gas for higher value uses (hard currency generating exports, vehicle fuels, chemical feedstocks, etc.). However, the coal industry experienced serious problems, including increasing depletion ratios and declining coal quality. Simply put, output in older fields, such as those in the Donets Basin, declined at an increasing rate relative to the expansion of new productive capacity in other fields.[9] Deteriorating geologic conditions in older fields were a major factor contributing to this situation. For example, production in the Donets Basin, which had been exploited since the 1700s, was plagued with increasing mine depth, thinning coal seams, and high methane concentrations.[10] Expansion of new productive capacity in eastern fields, notably the Kuznets, Ekibastuz, and Kansk-Achinsk basins, was fraught with a number of problems. Although the Kuznets still holds large reserves of high-quality coal, its development has suffered from labor shortages, inadequate investment, and insufficient transport capacity for delivery to consuming centers in the western USSR. Conversely, while coal formations in Ekibastuz and Kansk-Achinsk fields are close to the surface, allowing low-cost, non labor-intensive strip mining, these coals are very low in quality, and in Kansk-Achinsk they possess unusual chemical properties that create serious complications in transport and use.[11] Lower coal quality is not a problem restricted to Ekibastuz and Kansk-Achinsk coals; it appears to have been an ongoing, industrywide phenomena since the mid-1970s.[12] Declining coal quality constitutes a major problem for the industry's contribution to the energy sector. Declines in coal quality (expressed in kilocalories per unit weight) necessitate the consumption of a larger amounts of coal to produce the same amount of energy, ultimately placing greater strains on the extraction and transport infrastructure as well as on end use.

Oil did not become a dominant fuel source in the Soviet energy balance until the mid-1960s. Although oil expansion was impressive in the 1920s and 1930s, it did not compare with the success achieved in the coal industry. This reversed dramatically during the 1960s when oil production growth rates overtook coal production growth. The Soviet oil industry's impressive performance continued into the 9th (1971–1975) and 10th (1976–1980) Five-Year Plans. However, after 1980, oil production essentially stagnated at about 600 million metric tons a year.[13]

The Soviet oil industry, like coal, has experienced a gradual geographic shift in production to the east (Table 3). During the 1920s and 1930s, the Baku oil fields and the North Caucasus fields dominated oil production. However, the depletion of these fields, the discovery of the Volga oil fields in the late 1930s, and the German invasion in 1941 initiated a rapid expansion of production in the fields along the Volga and and in the Urals. The discovery of huge oil and gas reserves in West Siberia during the 1960s allowed the Soviets to continue rapid rates of expansion with a massive oil development program in West Siberia during the 1970s, just as production peaked

Table 3: Geographic Pattern of Conventional Primary Fuel Production, 1960-1990

Fuel	1960	1965	1970	1975	1980	1985	1990
Coal							
European USSR (%)	64.1	60.7	56.8	50.8	50.9	45.5	35.0
Eastern USSR (%)	35.9	39.3	43.2	49.2	49.1	54.5	65.0
Total tonnage[a]	510.0	578.0	624.0	701.0	716.4	726.4	703.0
Oil							
European USSR (%)	92.8	92.9	81.9	61.0	39.2	32.7	24.6
Eastern USSR (%)	7.2	7.1	18.1	39.0	60.8	67.3	75.4
Total tonnage	148.0	243.0	353.0	491.0	603.2	595.3	567.7
Gas							
European USSR (%)	98.9	84.8	67.5	50.4	37.3	21.8	13.3
Eastern USSR (%)	1.1	15.2	32.5	49.6	62.3	78.2	86.7
Total cubic meters[b]	45.3	128.0	198.0	289.3	435.2	642.9	814.7

[a] In million metric tons.
[b] In billion cubic meters.

Sources: A. M. Nekrasov, *Problemi razvitiye i razmeshcheniya toplivnikh baz SSSR,* pp. 10–11; Theodor Shabad,"News Notes," *Soviet Geography,* Vol. 19, No. 4 (April 1978), pp. 274, 278, 282; Matthew Sagers, "News Notes," *Soviet Geography,* Vol. 32, No. 4 (April 1991), pp.257, 266, 273.

in the Volga–Urals in 1974–1975. Thus the Soviets were able to maintain high growth rates of oil production and small depletion ratios by expanding production in new fields just as production in older fields began to decline. However, by the 1980s it had become obvious to the Soviets that their easily accessible oil reserves had been exhausted

As illustrated in Table 1, gas has become an important Soviet primary fuel only since the 1960s. Extremely rapid rates of growth were obtained during the 1960s, 1970s, and early 1980s. During the 1970s and 1980s, the timely development of the gas industry was something akin to a miracle for Soviet planners, offsetting failures in other fuel industries. Indeed, the Soviet gas industry was the only fuel industry to overfulfill plan output in the 10th (1976–1980) and 11th (1981–1985) Five-Year Plans.[14]

Like coal and oil, the Soviet gas industry has experienced an eastward shift in production, although over a much shorter period of time. Before the late 1950s, gas production was relatively insignificant and was centered in the Ukrainian and Volga–Urals oil and gas fields.[15] The rapid expansion of gas production that ensued in the 1960s took place in the Ukrainian, North Caucasus, and Cental Asian fields. As these peaked in the early 1970s (in the case of Central Asia, the early 1980s), the vast West Siberian oil and gas fields went into production. As a result of the massive efforts of the West Siberian gas development program, by 1980 West Siberian production accounted for more than 35% of Soviet gas production. During the 11th Five-Year Plan (1981–1985), almost the entire growth increment of gas production occurred in West Siberia, more than offsetting declines in European fields. Following up on these successes, Soviet plans envisioned an even greater role for gas in the country's energy balance beyond the 1990s. However, the spatial trends in Soviet gas development presented this industry with a difficult future.[16] Planners realized that the continued growth of production levels in the West Siberian fields after the late 1980s would necessitate the development of more northern fields (Yamburg and Yamal') in West Siberia. These fields presented much more difficult permafrost conditions and required complex technology and higher investment levels for successful exploitation.[17] Other more near-term constraints facing the gas industry were an insufficient gas distribution infrastructure and insufficient storage capacity.[18]

The above discussion illustrates the underlying trend for the Soviet energy sector—a geographic shift in the production of the three conventional primary fuels from West to East, away from the established demand centers. This geographic shift has resulted in increasing costs for these fuels and has presented an increasing strain on the investment resources of the national economy. Three trends are responsible for this: expanding extraction activities in less developed, inaccessible areas; declining output in old fields requiring increased investment to maintain net growth of production; and longer distances for fuel transport via pipelines and rail to consumers in the western USSR. These developments have resulted in price increases for all three of these fuels since 1965, and particularly since 1974.[19] Moreover, increasing the production of the three conventional primary fuels absorbed an increasing amount of capital investment. Capital investment per marginal increase of production for oil rose from 98 rubles per metric ton of standard fuel equivalent in 1965 to 290 rubles by 1980. Similarly, for gas the increase was from 47 rubles per ton of standard fuel equivalent to 165 rubles.[20] The three conventional primary fuel industries increased their share of total industrial investment from 19% during the 9th Five-Year Plan (1966–1970) to 26.5% during the 11th Five-Year Plan (1981–1985).[21] Some Western observers estimate even higher shares of total capital investment devoted to the fuel industries during the 11th and 12th Five-Year Plans.[22]

Development of the Electric Power Industry

A major consumer of conventional primary fuels since the 1940s has been electric power industry. Electric power is a secondary form of energy, converted from the burning of primary fossil fuels, nuclear fission (technically a primary fuel), or the use of ambient energy sources, such as waterpower or wind. Electricity is typically a more flexible and conveniently manipulated form of energy than the direct use of primary forms of energy. Moreover, electricity is a preferred form of energy because it frequently offers considerable gains in factor productivity and economies of scale.

The electric power industry was given high priority in the development of the Soviet economy. The Communist Party's highly publicized GOELRO (State Commission for the Electrification of Russia) project, formulated in 1920, was a clear manifestation of the party's commitment to the development of the electric power industry.[23] As can be seen in Table 4, the Soviet electric power industry has grown at a very fast pace since the 1920s. The vast majority of the expanding capacity in the Soviet power industry was based on thermal power stations, which relied on conventional primary fuels to generate electricity. In addition, the Soviets developed hydropower as a generating source (see Table 4). Since the early 1960s, however, hydropower's share of total capacity has stablized at just under one-fifth of total capacity. Hydropower's relative stagnation was due to the lack of suitable dam sites and competing interests of alternative power sources, particularly in the European USSR.

Spatially, the Soviet electric power industry has remained concentrated close to demand centers. In 1970, some 59.3% of Soviet electric-generating capacity was located in the European USSR, 70.0% if the Urals are included in this figure.[25] Within the European USSR, this capacity was concentrated in the Ukrainian SSR and the Volga, Central, and Northwest* economic regions, which together constituted more than 41% of the USSR's total generating capacity in 1970.[26] Soviet capacity additions since 1970 have remained centered in the European USSR, although this region's share of capacity appears to be gradually declining.[27]

The tremendous growth of thermal capacity resulted in an enormous increase in conventional primary fuel consumption. The use of consumption of conventional primary fuels at thermal power stations increased from 160 million tons of standard fuel in 1960 to 503 million

---

* This source included capacity in the Northern Economic Region in that reported for the Northwest Economic Region.

Table 4: Soviet Electric Power Capacity, 1940–1990

Capacity in thousands of Megawatts (MW) and percent share[a]						
Thermal		Hydro		Nuclear		
MW	%	MW	%	MW	%	Total MW
1940	9.6	85.8	1.6	14.2	—	—	11.2
1950	16.4	83.6	3.2	16.4	—	—	19.6
1955	31.2	83.9	6.0	16.1	—	—	37.2
1960	51.9	77.8	14.7	22.2	—	—	66.7
1965	92.8	80.7	22.2	19.3	—	—	115.0
1970	133.8	80.5	31.4	18.9	0.9	0.6	166.1
1975	172.1	79.2	40.5	18.6	4.9	2.2	217.5
1980	201.9	75.7	52.3	19.6	12.5	4.7	266.7
1985	225.1	71.4	61.7	19.6	28.3	9.0	315.1
1990	241.1	70.3	65.0	19.1	37.9	11.6	344.0

— less than .1
[a] figues rounded

Sources: Minenergo, *Razvitiye elektro-energeticheskogo khozyaystvo SSSR: khonologicheskiy ukazatel'* (Moscow: Energoatomizdat, 1987), selected years; A. A. Troitskii, "Energetika SSSR za 70 let," *Elektricheskiye stantsii,* No. 11, 1987, p. 3; Yu. N. Flakserman, *Teploenergetika SSSR 1921–1980,* p. 10; Goskomstat, *Narodnoye khozyaystvo v 1990,* p 395.

tons by 1980.[28] Equally important, thermal power station consumption as a share of total domestic conventional primary fuel consumption increased from approximately 23.8% in 1960 to 31.7% in 1980.[29] This significant increase in consumption occurred despite a somewhat impressive fuel savings program at thermal power stations during the 1950s, 1960s and 1970s. Efforts in fuel savings at thermal power stations have focused upon three engineering strategies: increasing the steam parameters in turbines (to improve energy conversion efficiencies), increasing unit scale, and expanding the application of cogeneration at electric power plants.[30] The heat rate,* a standard measure of fuel use for electric power generation declined from 581 grams per kilowatt hour (kWh) in 1951, to 367 in 1970, 328 in 1980 and 326 in 1985 and 1990.[31] As can be seen, despite the impressive gains in fuel savings at Soviet thermal power stations, over time these gains have declined noticably. Available evidence indicates that future gains in fuel savings at Soviet thermal power plants were unlikely to be significant.[32]

---

* The ratio of fuel consumption in grams of standard fuel to the output of electricity in kilowatt hours.

Another important trend in fuel use at Soviet thermal power stations was the shift from solid fuels, especially coal and peat, to high-quality fuels, such as oil and gas (see Table 5). This transition mirrored the gradual switch from solid to liquid fuels in primary fuel extraction. While this switch to more efficient burning liquid fuels facilitated overall energy savings in terms of reducing the heat rate at thermal stations, for the economy as a whole it had negative consequences. The opportunity costs (costs in terms of revenue foregone through more profitable use of the resource) of oil and gas use at electric power stations were considerable. This is especially true for oil, as other high-value users, such as agricultural machinery, automobile and truck transport, and especially hard currency exports compete for oil.[33]

Table 5: Fuel Mix For Thermal Power Stations, 1940–1980

	Percent share			
	Coal	Oil	Gas	Peat[a]
1940	69.0	8.7	2.3	20.0
1960	70.9	9.3	12.3	7.5
1970	46.1	22.5	26.0	5.4
1975	41.3	28.8	25.7	4.5
1980	42.5	28.0	25.1	4.4

[a] Although reported as peat may include wood as well.

Source: Flakserman, p. 168.

Thus, within the electric power industry there was a spatially dichotomous trend between fuel supply and consumption. The extraction of the three major primary fuels shifted to the east since the 1960s, while electric power capacity continued to expand rapidly in the European USSR. This trend led to higher costs of electricity production in the European USSR and placed an increasing strain on the Soviet rail transport system. Despite the development of hydroelectric resources and a moderately successful fuel savings program for thermal power stations, the Soviet electric power industry has remained largely dependent upon conventional primary fuels; and over time, its share of total conventional primary fuel consumption increased. This came at a time when greater emphasis was being placed on the development of competing uses for conventional primary fuels, particularly oil and gas. These competing uses included a growing chemical industry, automobile transport, and perhaps most importantly, especially after 1973, the demands of rapidly expanding oil and gas exports to Eastern Europe and the West.

Heat Supply

The production of heat for industrial, municipal, and residential use was another major consumer area within the Soviet energy sector. In the USSR, heat was supplied from several sources. These included "centralized" sources—thermal cogeneration stations (TETs) and large utility boilers (typically with capacities greater than 20 gigacalories [Gcal] per hour).* These centralized sources supply heat through district heating systems to surrounding industrial and municipal facilities and residences. Noncentralized sources include small boilers or heaters in individual dwellings, apartment flats, or individual factories.

A thorough description of trends in Soviet heat supply is difficult, because Soviet sources vary considerably in reporting data concerning heat production, categorization of heat producers, and even unit of measurement (this study uses the calorie as the unit of heat production).[34] For example, estimates of heat supply from centralized sources range from 1,303 million Gcal (reportedly 76.4% of total heat production) to 1,550 million Gcal (reportedly 59.5% of total heat production) in 1980.[35] Nevertheless, according to Melent'yev and Markov, large and small boilers as well as individual unit heaters intended specifically for heat production accounted for 24.5% of primary fuel consumption in 1960, declining slightly to 22.3% by 1980 (see Table 6).[36]

This relative decline in heating's share of total fuel consumption was largely due to the efficiencies gained through the centralization of heat supply for urban and industrial areas. Over time, the Soviet planners attemped to lower fuel use at inefficient small boiler and heater units by increasing the role of centralized sources that distribute heat through district heating systems. Centralized sources for district heating systems include TETs and large or regional boilers.

As can be seen in Table 7, centralized sources have expanded

---

* Centralized sources also include heat from district thermal power stations (GRES) and nuclear power stations (AES). These types of stations while designed specifically for power production, bleed excess steam from the turbines; the heat from this is then used for local district heating systems.

**Table 6: Fuel Consumption for Thermal Electric Power
Generation and Heat Production, 1960–1980**

	Millions of tons standard fuel[a] and percent share		
	1960	1970	1980
Thermal power stations	160.0 (23.8)	321.0 (28.7)	503.0 (31.7)
Large boilers	60.0 (8.9)	101.0 (9.0)	167.0 (10.5)
Small boilers and individual units[b]	105.0 (15.6)	145.0 (13.0)	188.0 (11.8)
Total conventional primary fuel consumption	673.0	1,118.0	1,587.0

[a] One kilogram of standard fuel equals 7,000 kilocalories.
[b] Boiler capacities below 20 Gcal per hour.

Sources: Station and boiler data from L. A. Melent'yev and A. A. Markov, *Energeticheskii
kompleks SSSR*, 1983, p. 44; USSR fuel consumption from Melent'ev and Markov, Table 1.4, p.
43, minus primary electricity as reported in Table 1.2, p. 40.

since the 1960s. Spatially, district heating was concentrated in the relatively highly urbanized European USSR. Although comprehensive regional data are scarce, one source indicates that in 1970, 68.2% of Soviet heat use from centralized sources occurred in the European USSR, 78.9% if the Urals are included in this figure.[37]

Nuclear Power as an Alternative Energy Source

Given the combination of circumstances developing in the Soviet energy sector since the 1960s, Soviet planners and policy-makers turned to nuclear energy as an alternative source of power for the electric power industry and to a much lesser extent heat production. The Soviet Union was the first country in the world to operate a commercial power reactor, the 5 megawatt (MW) Obninsk reactor, which began operating in June 1954.[38] This and other successful pilot projects involving nuclear electric power generation convinced Soviet researchers of the viability of using nuclear sources for electrical power and heat generation.

With the relative decline beginning in the late 1960s, in con-

Table 7: Soviet Heat Production By Source, 1965–1980

	Millions of Gcal and percent share							
	1965		1970		1975		1980	
Centralized Sources								
TETs[a]	462.0	(29.8)	688.0	(32.4)	920.0	(34.7)	1,160.0	(35.0)
Large boilers[b]	44.0	(2.8)	139.0	(6.5)	290.0	(10.9)	450.0	(14.1)
Secondary sources[c]	31.0	(2.0)	45.0	(2.1)	85.0	(3.2)	115.0	(3.6)
Local Sources								
Small boilers	464.0	(29.9)	570.0	(26.8)	620.0	(23.4)	690.0	(21.6)
Individual units	549.0	(35.5)	682.0	(32.1)	735.0	(27.7)	775.0	(24.2)
Total	1,550.0		2,125.0		2,650.0		3,200.0	

[a] Although listed as TETs may also include a small amount of heat from GRES and AES.
[b] Defined as heat capacity greater than 50 Gcal/h.
[c] Alternatively known as heat utilization units (*utilizatsionniye ustanovkii*), heat recycled from industrial furnaces, kilns, etc.

Sources: Melent'yev and Markov, p. 49; also see Flakserman, p. 167, who reports similar figures for centralized sources but very different figures for localized sources, apparently because he cites consumption rather than production as Melent'yev and Markov do.

ventional fuel production of coal, oil, and gas in the European USSR, fuel costs for power station use began to rise. It was within this context that nuclear power became an attractive alternative to conventional fuels. Starting in the late 1960s, several studies by Soviet research institutes indicated that nuclear power was competitive with other fuels for power generation from a cost standpoint in the European USSR.[39] Other studies suggested nuclear power as a competitive alternative in the Far North.[40] As is discussed later, these studies made important assumptions about the use of the technology, including facility size and reactor safety, that lowered estimated capital costs and heavily influenced siting and safety policies in the Soviet nuclear power industry.

Since 1971, when a formal decision was announced to expand nuclear power-generating capacity, the Soviet nuclear power industry expanded rapidly.[41] Although the Soviet Union's extremely ambitious plans for the nuclear industry were not fulfilled, the Soviet nuclear industry's growth was nevertheless, impressive (see Table 8). Not surprisingly, nuclear capacity came to play a crucial role in power capacity additions in the European USSR by the 1980s (the 11th

Table 8: Expansion of Soviet Nuclear Capacity, 1971–1990

| | Capacity additions in MW | | | | % Completed total |
	Planned nuclear	Completed nuclear	Planned total	Completed total	that is nuclear
1971–1975	7,000 (a)	3,765 (b)	65,000–67,000 (c)	58,100 (d)	6.5
1976–1980	13,000	7,825 (b)	67,000–70,000 (e)	53,968 (d)	14.5
1981–1985	25,000 (f)	15,618 (b)	68,900 (g)	51,175 (d)	30.5
1986–1990	40,500 (h)	11,500 (b)	85,300 (h)	35,700 (i)	32.2

Sources: (a) Ryl'skiy, p. 31.
(b) Compiled from IAEA, *Operating Experience with Nuclear Power Stations in Member States in 1987* (Vienna: IAEA, 1988); and Goskomstat, *Promishlennost' SSSR*, p. 134.
(c) "The Directives of the 9th Five-Year Plan," *CDSP*, Vol. 23, No. 16 (May 14, 1971), from *Izvestiya*, April 11, 1971, p.1.
(d) Goskomstat,*Kapital'noye stroitelstvo: statisticheskiy sbornikh*, (Moscow: Finansy i Statistika, 1988), p. 53.
(e) "Guidelines for the 10th Five Year Plan-II," *CDSP*, Vol. 28, No. 16 (May 16, 1976), from *Izvestiya*, March 7, 1976, pp.2–8.
(f) "Guidelines for the 11th Five-Year Plan," *CDSP*, Vol. 33, No. 16 (May 20, 1981) p. 11, from *Izvestiya*, March 5, 1981, p. 3.
(g) Nekrasov and Troitskiy, p. 43.
(h) Troitskiy, *Energetika SSSR v 1986 –1990 godakh* (Moscow: Energoatomizdat, 1987), p. 173.
(i) Mathew Sager, "News Notes," *Soviet Geography*, Vol. 32, No. 4 (April 1991), p. 281 who sites *Ekonomika i Zhizn'*, No. 5, January 1990.

and 12th Five-Year Plans) as Table 9 illustrates. For the entire country, nuclear energy became an important generating source for the electric power industry, accounting for approximately 10% of capacity and production by 1985 (see Tables 10 and 11).

The use of nuclear energy for heat production also was pursued by the Soviet Union, although at a much more modest scale than for the generation of electricity. Total heat production from nuclear sources was reported as 1.8 million Gcal in 1980 and 2.6 million Gcal in 1985.[42] While the extent of the application of nuclear energy for heat production in the Soviet Union exceeded that of any other country, nuclear heat production in the USSR amounted to less than 1% of total heat production in 1985.[43]

Thus, despite the USSR's favorable energy resource endowments, planners and policy-makers during the Soviet period were faced with an increasing difficult set of choices in managing the country's energy sector. As primary fuels became depleted in the western fields of the USSR, primary fuel extraction shifted eastward.

Table 9: Capacity Additions in the European USSR, 1971-1987

| | Capacity in MW | | |
	Total completed	Completed nuclear	% Total that is nuclear
1971–1975	42,107	3765	8.9
1976–1980	35,265	7825	22.2
1981–1985	32,062.5	15,618	48.7
1986–1987	12,816	4,500	35.1

Sources: Goskomstat, *Kapital'noye stroitelstvo*, p. 53.

Table 10: Installed Capacity at Soviet Electric Power Stations, 1950-1990

| | Gross capacity in MW | | | | |
	Thermal	Hydro	Nuclear[a]	Total	% Nuclear[b]
1950	16,396	3,218	—	19,614	—
1955	31,250	5,996	—	37,246	—
1960	51,940	14,781	—	66,721	—
1965	92,789	22,244	310	115,033	0.3
1970	134,782	31,368	875	166,150	0.5
1975	176,969	40,515	4,671	217,484	2.1
1980	214,446	52,311	12,492	266,757	4.7
1985	253,164	61,724	28,110	314,888	8.9
1987	269,571	62,695	34,400	332,266	10.4
1989	244,200	64.400	37,400	341,000	10.9
1990	241,100	65,000	37,500	344,000	11.6

[a] Does not include pilot or prototype reactors. These include reactors at Obninsk, Dimitrovgrad, and Troitsk (also known as the "Siberian" station) in the southern Urals. Together these reactors constitue an estimated 672 Mw of capacity, although it is unlikely they were all in operation at any one time.

[b] Percent of total capacity reported as nuclear for that year.

Sources: Data for 1950–1960 from Minenergo, *Razvitiye elektro-energeticheskogo khozyaistva*, pp. 60, 67, 78; 1960–1987 data from *Promyshlennost' SSSR*, p. 134; 1989 and 1990 data from "Plan GOELRO i razvitiye energetiki," *Elektricheskiye stanstii*, No. 12 (December 1990), p. 15 and Goskomstat, *Narodnoye Kozyaystvo SSSR v 1990*, p. 395.

Table 11: Soviet Electric Power Production, 1950–1990

| | Millions of kWh[a] | | | |
	Thermal[b]	Hydro	Nuclear	Total	% Nuclear[c]
1950	78,535	12,691	—	91,226	—
1955	147,060	23,165	—	170,225	—
1960	241,362	50,913	—	292,275	—
1965	425,238	81,434	1,410	506,672	0.3
1970	616,549	124,377	3,499	740,926	0.5
1975	912,620	125,987	20,266	1,038,607	2.0
1980	1,037,066	183,889	72,923	1,293,878	5.6
1985	1,162,186	214,530	167,401	1,544,117	10.8
1987	1,258,115	219,825	186,984	1,664,924	11.2
1990	1,283,500	233,000	211,500	1,728,000	12.2

— less than 0.1

[a] Production during the year cited.

[b] These figures are from *Promyshlennost' SSR* which, from the sources available, provides the most detailed figures for production over the period of interest. However, this source erroneously includes nuclear production among that produced at thermal stations. The author has subtracted reported hydroelectric and nuclear production from the total to generate thermal figures. These thermal figures are consistent with rounded figures from other sources. Apparently, the reported production from thermal sources also includes production from unconventional sources such as geothermal and wind, which according to one source (Goskomstat, *Material'no-tekhnicheskoye obespecheniye*, p. 78) amounted to 600 million kWh of reported production in the years 1975, 1980, and 1985.

[c] Percent of total electricity production reported as nuclear for that year.

Sources: Data for 1950–1960 from Minenergo, *Razvitiye elektro-energeticheskogo khozyaistva*, pp. 60, 67, 78; 1960–1987 data from *Promyshlennost' SSR*, p. 134; 1990 data from Matthew Sagers, "News Notes," *Soviet Geography*, Vol. 32, No. 4 (April 1991), p. 281, who cites *Ekonomika i zhizn'*, No. 5, January 1990.

This eastward shift occurred for coal, oil, and gas in the 1960s and 1970s and placed increasing burdens on the Soviet economy. Over the same time, that is the late 1950s, 1960s, and 1970s, the Soviets greatly expanded electric power capacity. Most of this expansion was based on thermal capacity that relied heavily on solid fuels such as coal and peat until the 1960s when a sizable switch to oil and gas occurred. In attempting to avoid the opportunity cost burdens that associated with such heavy reliance on high-value liquid fuels and increasingly expensive coal, Soviet planners looked toward forms of primary electricity such as hydropower and nuclear power. Hydropower, availablity however, was limited in the European USSR. Within this context, Soviet planners saw nuclear power as a panacea to many of their energy problems. In 1971 the Party embarked upon an ambitious program to rapidly introduce nuclear electric power in the European USSR. By the early 1980s, less ambitious but nevertheless substantial plans were introduced for nuclear heat generation.

Notes

1. Estimated by the Central Intelligence Agency (CIA) at year-end in 1974. CIA, *CIA Handbook of Economic Statistics* (Washington, D.C.: Central Intelligence Agency, 1975), p. 68. For discussion of Soviet perceptions of their own reserves during this period, see Chung, pp. 64–70.

2. As of year-end 1989, Soviet explored natural gas reserves amounted to 1,450 trillion cubic feet; coal reserves amounted to 182,000 million metric tons (second only to the United States, Soviet unexplored reserves are estimated at 680,000 million metric tons although this figure is very tentative), Soviet oil reserves were treated as a state secret, however Western estimates have placed them between 50 and 80 billion barrels, second only to Saudi Arabia, Iraq, and Kuwait. CIA, *USSR Energy Atlas* (Washington, D.C.: CIA, 1985), p. 34; CIA, *CIA Handbook of Economic Statistics* (Washington D.C.: Central Intelligence Agency, 1990), p. 68.

3. The European USSR's share of the total population dropped (1981, 69.4%) largely due to the high natural growth rates of Central Asia's population. Goskomstat, *Naseleniye SSSR 1987* (Moscow: Finansy i Statistika, 1988), pp. 16–31.

4. A. S. Pavlenko and A. M. Nekrasov (eds.), *Energetika v SSSR 1971–1975 godakh* (Moscow: Energiya, 1972), p. 79.

5. CIA, *USSR Energy Atlas*, p. 35.

6. Pavlenko and Nekrasov, p. 147.

7. Tsentralnoye Statistichskoye Upravleniye (TsSU), *Strana Sovetov za 50 let* (Moscow: Statistika, 1967), p. 68.

8. While Soviet coal production reached an all-time high in 1985 during the 12th Five-Year Plan at 726 million tons, a 2% increase in output over 1980 levels, this compares poorly with past growth rates as well as the expectations of planners. Matthew Sagers, "News Notes," *Soviet Geography*, Vol. 27, No. 4 (April, 1986), p. 286; also see Ed Hewett, *Reforming the Soviet Economy: Economy versus Efficiency* (Washington, D.C.: The Brookings Institution, 1988), p. 86.

9. For a more detailed discussion,see Hewett, pp. 85–88.

10. Soviet mines in the Donets Basin have an average depth of 605 meters (eight times the average for U.S. mines) and seam thicknesses average less than one meter, requiring special equipment and extraction procedures. David Warner and Louis Kaiser, "Development of the USSR's Eastern Coal Basins," *Gorbachev's Economic Plans*, Study paper submitted to the Joint Economic Committee, U.S. Congress, November 23, 1987 (Washington, D.C.: U.S. Government Printing Office, 1987), p. 534.

11. Ibid.

12. Rising ash content in Soviet coal is another negative quality trend. A. M. Nekrasov and A. A. Troitskii (eds.), *Energetika v SSSR 1981–1985* (Moscow: Atomenergoizdat, 1982), p. 224; K. F. Raddatis and K. V. Shakhsuvarov, "O poteryakh v narodnom khozyaistve iz-za ponizhennogo kachestva uglei dlya teplovikh elektrostanstii," *Elektricheskiye stantsii*, No. 1 (January 1985), pp. 6–10.

13. Ed Hewett, *Energy Economics and Foreign Policy in the Soviet Union* (Washington, D.C.: The Brookings Institution, 1984), p. 65, and Theodore Shabad, "News Notes," *Soviet Geography*, Vol. 27, No. 4 (April 1986), p. 259.

14. Matthew Sagers, "News Notes," *Soviet Geography*, Vol. 32, No. 4 (April 1991), p. 253.

15. Hewett, *Energy Economics*, p. 67.

16. M. G. Krulov et al., "Prioritetniye napravleniya i gosudarstvenniye programmi nauchno-tekhnicheskogo progressa v

proizvodstve i ispol'zovanii energeticheskikh resursov," *Teploenergetika*, Vol. 36, No. 1 (January 1989), pp. 2–6.

17. For more detailed discussion, see Alrid Moe and Helge Ole Bergson, "The Soviet Gas Sector: Challenges Ahead," in *The Soviet Economy: A New Course* (Brussels: NATO, 1987), pp. 153–194.

18. L. A. Melent'yev and A. A. Makarov, *Energeticheskiy kompleks SSSR* (Moscow: Economika, 1983), p. 123.

19. Matthew Sagers and Albina Tretyakova, "Constraints in Gas and Oil Substitution in the USSR: The Oil Refining Industry and Gas Storage," *Soviet Economy*, Vol. 2, No. 1 (January–March 1986), pp. 72–81.

20. Ibid. p. 73.

21. TsSU, *Narodnoye khozyaystvo SSSR, v 1980*, p. 336; TsSU, *Narodnoye khozyaystvo SSSR, v 1985*, p. 368.

22. See Leslie Dienes, "The Energy System and Economic Imbalances in the USSR," *Soviet Economy*, Vol. 1, No 4 (September–December 1985), pp. 340–372; Albina Tretyakova and Barry Kostinsky, "Fuel Use and Conservation in the Soviet Union: The Transportation Sector," in *Gorbachev's Economic Plans*, Study paper submitted to the Joint Economic Committee, U.S. Congress, November 23, 1987 (Washington, D.C.: U.S. Government Printing Office, 1987), pp. 544–565.

23. "Plan GOELRO i razvitiye energetiki," *Elektricheskiye stantsii*, No. 12 (December 1990), pp. 13–15.

24. See Thane Gustafson, *Reform in Soviet Politics: Lessons of Recent Policies on Land and Water* (Cambridge, MA: Cambridge University Press, 1981), Chapters 3 and 7; Pavlenko and Nekrasov, pp. 143–147.

25. V. A. Ryl'skiy, *Elektroenergeticheskaya baza ekonomicheskikh rayonov SSSR* (Moscow: Nauka, 1974), pp. 109–125, 201–215.

26. Ibid.

27. The European USSR and the Urals accounted for 67.8% of capacity additions during the 9th Five-Year Plan (1971–1975) and 65.3% during the 10th Five-Year Plan (1976–1980). Goskomstat, *Kapital'noye stroitel'stvo: statisticheskiy sbornikh* (Moscow: Finansy i Statistika, 1988), p. 53.

28. Melent'yev and Makarov, p. 44.

29. Based on data from Melent'yev and Markov, pp. 40–43.

30. Fuel savings at electric power stations has been fairly successful, largely because this is a major performance indicator for the Ministry of Power and Electrification (*Minenergo*). Increases in the steam parameters of turbines (higher steam temperatures and pressures) and unit sizes (boilers and turbines) were intended to generate improvements in the heat rate. Increasing the application of cogeneration improves the efficiency of energy use as waste heat from turbine steam is recycled and used in district heating of urban areas and industrial facilities. Soviet capacity for thermal cogeneration stations (TETs) increased from 16,800 MW in 1960 to 75,500 MW in 1980, the latter figure constituting some 37.3% of total thermal capacity for that year. Yu. N. Flakserman, *Teploenergetika SSSR 1921–1980* (Moscow: Nauka, 1981), pp. 103–108; V. I. Gorin and Ya. F. Masalov, "Povysheniye effektivnosti teplovikh elektrostanstiy," *Teploenergetika*, Vol. 28, No. 6 (June 1981), pp. 5–8; "Plan GOELRO i razvitiye energetiki," p. 15.

31. Flakserman, pp. 116, 153; and TsSU, *Narodnoye khozyaystvo SSSR, 1980*, p. 155.

32. Hewett, *Energy Economics*, pp. 114–115; Gorin and Masalov, pp. 5–8.

33. Total consumption of liquid fuels from the transportation sector increased from 68.3 million tons of standard fuel equivalent in 1970 to 110.8 million tons in 1984, an increase of some 162%, Tretyakova and Kostinsky, p. 553. Between 1970 and 1980, net exports of oil (primarily to Eastern Europe and the West) increased from 94 million tons of standard fuel equivalent to 165.2 million tons, an increase of some 176%. The increase for natural gas was even more dramatic, from net imports of some 0.3 million tons of standard fuel equivalent in 1970, to net exports

of some 60.2 million tons in 1980. Goskomstat, *Material'no-tekhnicheskoye obespecheniye narodnogo khozyaystva SSSR* (Moscow: Finansy i Statistika, 1988), pp. 76–77.

34. This lack of consistency is no doubt due to the complete lack of centralization and adherence to formal arrangements in relations between heat consumers and producers, with the result of rampant falsification by both producers and consumers. For a more detailed discussion, see Boris Yudzon, "Decision Making in the Soviet Heat and Fuel Supply Systems: Contradiction in Consumer Supplier Relations," in Katherine Young, pp. 59–93.

35. Flakserman, p. 187; Nekrasov and Troitskiy, p. 73.

36. Melent'yev and Makarov, pp. 48–50, 122–123.

37. Pavlenko and Nekrasov, p. 78.

38. Minenergo, *Razvitiye elektroenergeticheskogo khozyaystva SSSR: khronologicheskiy ukazatel'* (Moscow: Energoatomizdat, 1987), p. 65.

39. William Kelly et al., "The Economics of Nuclear Power," *Soviet Studies*, Vol. 34, No. 1 (January 1982), pp. 58–59; Yu. I. Koryakin, "Mathematical Modeling of the Developing Nuclear Power Generation," *Soviet Atomic Energy*, Vol. 36, No. 6 (December 1974), pp. 586–591 (from *Atomnaya Energiya*, Vol. 36, No. 6, June 1974, pp. 419–422); D. G. Zhimerin, "The Present and Future of the Soviet Power Industry," *Thermal Power Engineering*, Vol. 17, No. 3 (March 1970), p. 5 (from *Teploenergetika*, Vol. 17, No. 3, March 1970, p. 4).

40. Ibid.

41. "The Directives of the Five-Year Plan," *CDSP*, Vol. 23, No. 18 (May 8, 1971), p. 13 (from *Izvestiya*, April 11, 1971, p. 1).

42. Goskomstat, *Material'no-tekhnicheskoye obespecheniye*, p. 79.

43. Ibid.

Chapter III

The Institutional Setting in the Soviet Nuclear Power Industry, Early Years to 1986

As pointed out in Chapter I, several researchers have argued that the institutional and procedural environment that exists in a particular country for a particular industry has a major influence on the scale and pace of adoption of a hazardous technology such as nuclear-based power generation. Nelkin and Pollak as well as Kitschelt advance institutional models to explain differences in the development of nuclear power in several countries.[1] This chapter identifies and describes the participants and institutional setting for locational decision-making for nuclear power facilities in the Soviet Union during the period between 1958 and 1986. This period, encompasses the early years of the Soviet nuclear industry's expansion up to the Chernobyl' accident.

Institutions Responsible for Energy Policy and Planning

Over time, a plethora of organizations has been involved in overall energy policy-making, policy in the nuclear industry, and individual site selection for power plants in the Soviet Union. The foremost characteristic of Soviet industrial decision-making was its hierarchical nature and extreme centralization. The planning and management in any industry in the former Soviet Union can best be conceptualized in four distinct functional hierarchical levels. These levels included: (1) policy-making institutions, (2) central planning institutions, (3) branch planning and management institutions, and (4) organizations and enterprises directly engaged in research, development, and production.[2] Institutions at all four levels were involved in decision-making in the nuclear industry and, directly or indirectly, in individual project making and the siting issue.

During the period discussed, basic energy policy and planning decisions pertaining to the allocation and application of energy technology on a national and regional scale were carried out by the highest levels of the Communist Party of the Soviet Union (CPSU) and the government or state hierarchy. The CPSU maintained supreme policy-making authority in the Soviet Union. The party exercised control over the basic directions of the economy—sectoral

and regional investment, general trends in foreign economic
relations, large-scale regional development projects, etc. Within the
CPSU, the *Politburo* was considered the highest policy-making
authority. The *Politburo* was supported by the Secretariat of the
Central Committee of the CPSU, whose departments oversaw party
and government activities and provided the *Politburo* with expert
advice and policy proposals (see Figure 1).[3]

In the Soviet Union, the upper levels of the government or state
hierarchy also played a significant role in policy formulation. Hewett
describes the relationship between the party and government in
formulating nuclear policy as follows:

> the party decides on the general policy regarding the
> development of nuclear power in the USSR; whereas the
> government, having provided much of the supporting technical
> and economic information, must implement the decision and
> deal with the practical task of mixing nuclear with other sources
> of energy in order to meet the economy's energy needs.[4]

Occupying the highest level of the state hierarchy was the USSR
Council of Ministers. Within the USSR Council of Ministers was the
organizational entity responsible for central planning—the State
Planning Committee of the USSR (*Gosplan*). With respect to energy
policy, *Gosplan* coordinated policy formulated from above with the
production capabilities of lower level production organizations in the
state hierarchy and incorporated them into long-term, five-year, and
annual plans.[5] These plans thus reflected the broad policy
guidelines established by the party and guided lower level branch
planning and management institutions and ministries in their
respective activities. Ultimately, *Gosplan* not only coordinated the
formulation of these plans, but supervised their enactment and
monitored their fulfillment.

Decisions made by these policy and planning organizations
were facilitated by the efforts of a vast array of research and advisory
organizations. The *Politburo* of the CPSU was supported by the Party
Secretariat, which conducted policy research. Similarly, the USSR
Council of Ministers was supported by the *Referentura*, an advisory
committee that provided research on requested topics including
energy.[6] The USSR Council of Ministers and *Gosplan* also received
key input from other research and advisory institutions. In the
realm of regional energy and nuclear power, these institutions
included the USSR State Committee for Science and Technology
(GKNT), the USSR Academy of Sciences (AN SSSR) and its special
Siberian branch (SO AN SSSR), the USSR State Committee on the

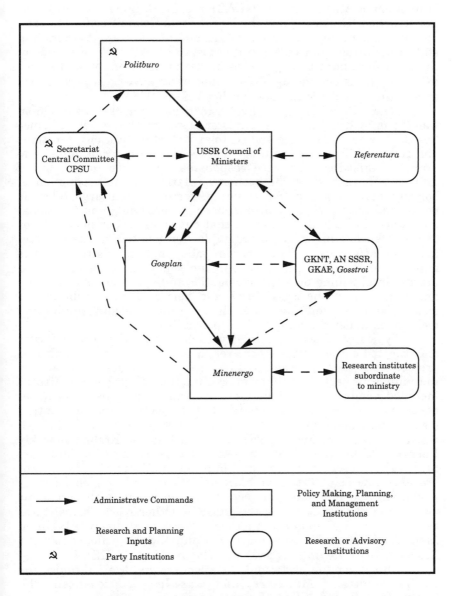

Figure 1: Institutions Involved in Policy-Making and Planning in the Soviet Nuclear Power Industry, 1970-1986

Utilization of Atomic Energy (GKAE), the USSR State Committee for Construction (*Gosstroy*), and the USSR State Committee for Material and Technical Supply (*Gossnab*).[7] In addition to these coordinating state committees and research organizations, research institutes of the individual ministries provided important information and advice in energy decision-making.[8] This appears to have been particularly true in the case of nuclear power policy before 1986.[9]

Next in the hierarchy of state institutions in the Soviet Union was the ministry. The Soviet ministerial system that developed since the 1930s was an intricate and complex web of institutions, functions and responsibilities. Essentially there were two types of ministries: branch ministries that were responsible for the management of different "branches" of the economy relating to the production of boardly defined goods (for example, ferrous metallurgy, electric power and tractor or agricultural machine building);, and functional ministries that manage activities that cut across the responsibilities of the branch ministries (for example, foreign trade, finance, and internal affairs).[10] While not a policy-making institution, the ministry had enormous influence in the policy and decision-making process in providing detailed information to high-level policy-makers and planners on the capabilities and performance of subordinate enterprises as well as research on products and production techniques under the ministry's responsibility.

Typically, a branch ministry such as the Ministry of Power and Electrification (*Minenergo*) relied on large basic research and project institutes (*proyektiye instituti*) to evaluate regional and individual projects. The primary function of project institutes included conducting engineering-cost studies, preparing technical documents necessary for investment projects, and formulating development plans for regions and branches of industry.[11] Indeed, research institutes under *Minenergo*, such as the Krzhizhanovskiy Energy Institute (ENIN) and project institutes such as the All-Union State Thermal Power Station Planning Institute (*Teploelektro-proyekt*), conducted extensive basic design research, regional energy modeling, and individual project studies to evaluate the economic viability of nuclear power in the USSR.[12] Other research institutes subordinate to *Minenergo* include, (the Dzerzhinskiy All-Union Institute of Heat Engineering (VTI) and the All-Union Scientific Research Institute of Electric Power Engineering (VNIIE).[13]

Altogether the party and state leadership—the *Politburo*, the USSR Council of Ministers, and *Gosplan*—represented the institutions that fulfilled the role of national policy-makers and planners. These policy-makers and planners, in turn, relied heavily on information from high-level advisory institutions such as the AN SSSR and the GKNT, the individual ministries below them, as well as research and project institutes subordinate to those ministries. Because the Soviet Union's economy was managed essentially through a system of administrative commands in the form of planning targets and directives sent from higher level entities in the organizational hierarchy to lower ones, national policy-makers had enormous influence in technological choice, the extent of its adoption, and to a limited degree the spatial character of its application on a regional level.

Institutions Involved in Site Selection

Actual siting and project decisions were, however, determined by the branch ministry. The *Minenergo* was, for the period under discussion, the ministry responsible for the siting, construction, and operation of electric power stations, including nuclear power stations. Between 1965 and 1986, *Minenergo* was a centralized national utility responsible for power station planning, construction, and operation as well as grid operation and the distribution of electric power.*[14] Historically, *Minenergo* operated the vast majority of the Soviet Union's electric power capacity (89% as of 1980; 91% as of 1987), the remainder being provided by small industrial stations under the management of other ministries.[15]

The existence of a single national utility or power ministry such as *Minenergo* centralizes decision-making, and in theory this would facilitate the study of site selection and project formulation. Unfortunately in the case of the USSR, considerable secrecy surrounded locational decision-making at the ministerial level. Secrecy definitely surrounded decision-making in the Soviet nuclear power industry, a state of affairs that has attracted considerable public attention since the Chernobyl' accident.[16]

Despite the lack of information, it appears that within *Minenergo* several organizations were involved in the facility site selection process (see Figure 2). In the design and construction of a

---

* Although as is discussed later, since the Chernobyl' accident the ministerial organization for power station operation has changed. Between 1953 and 1965, the Soviet electric power industry went through several organizational changes, which at times included the formal separation of power station construction and operation and between 1962 and 1963 separate republican ministries in the larger republics. Between 1963 and 1965 the Soviet electric power industry was known as the USSR State Industrial Committee for Power and Electrification.

nuclear power facility, *Minenergo* would delegate one of its subordinate design institutes to coordinate and supervise project work, which included the selection of a site. These design institutes were responsible for developing the project design, collecting the necessary technical data and documentation, and coordinating the activities of all organizations involved in the construction of the facility.[17] It is known that until 1986 the vast majority of projects were delegated to one of two design agencies according to the type of reactor being installed. These institutes were the All-Union Hydroproject Planning, Surveying, and Research Institute (*Gidroproyekt*), which was responsible for plants with RBMK-type reactors (*reaktor bol'shoy mozhnosti kanal'nyye*—graphite-moderated, water-cooled reactor), and *Teploelektroproyekt*, responsible for plants with VVER-type reactors (*vodo-vodyanoy energeticheskiy reaktor*—pressurized-water reactor).[18]

During the initial stage of project development, design drafts or project technical assignments would be formulated by the relevant design institute. It was during this stage that alternative sites were evaluated and a final site selected. Before the final submission of the design draft a number of different agencies and organizations would be consulted including the relevant construction agencies, such as the Main Administration for Atomic Power Station Construction (*Glavatomenergostroy*) and Administration for the Construction of Atomic Power Stations (*Atomenergostroy*) both under *Minenergo*, as well as agencies responsible for health, safety, and environmental protection.[19] Before 1983 these agencies included the State Supervisor for Sanitation of the Ministry of Health (*Gossannadzor*), the USSR State Committee of Standards (*Gosstandart*), the USSR State Technical Inspectorate (*Gostekhnadzor*), and the USSR State Inspectorate of Fisheries (*Rybnadzor*).* After input from these agencies was received, the design draft would be further modified by equipment suppliers—the Ministry of Power Machine Building (*Minenergomash*) and the Ministry of Medium Machine Building (*Minsredmash*), endorsed by *Minenergo*, and then submitted to *Gosstroy* for final approval.[20] At this point, *Gosstroy* would evaluate the design draft according to economic criteria and suggest changes that were usually intended to reduce construction costs. After the endorsement of the design draft, work would proceed on to the technical project in which specifications for the facility and equipment were worked out. Typically, site preparation and

---

* After 1983 *Gossannadzor* and *Gosnadzor* were replaced by the USSR State Committee on the Supervision of of Nuclear Power Safety (*Gosatomenergonadzor*), see p. 66.

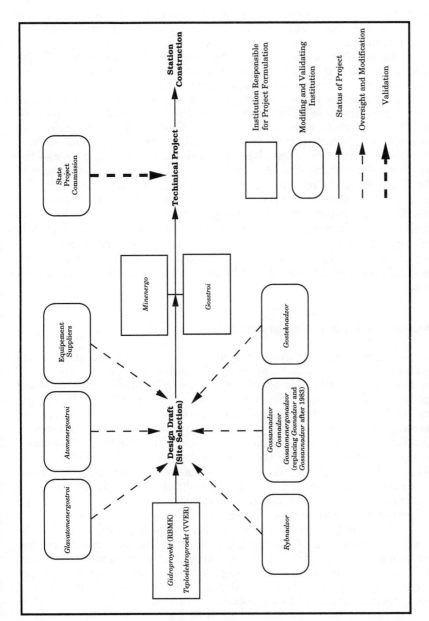

Figure 2: Institutional Framework for Siting Nuclear Power Stations

Station Construction

Technical Project

State Project Commission

Minenergo

Gosstroi

Equipement Suppliers

Atomenergostroi

Glavatomenergostroi

Gidroproyekt (RBMK) Teploelektroproekt (VVER)

Design Draft (Site Selection)

Gosatomenergonadzor Gosnadzor Gossannadzor (replacing Gosnadzor and Gossannadzor after 1983)

Gosteknadzor

Rybnadzor

Institution Responsible for Project Formulation

Modifing and Validating Institution

Status of Project

Oversight and Modification

Validation

development would begin after design draft endorsement. Actual plant construction would not begin until after the technical project was completed.

As mentioned earlier, a formal public licensing procedure for validating a planned facility did not exist in the Soviet Union during this period. The only process approximating a Western licensing procedure was the validation process during the review of the design draft before its final submission, as well as a final project review by the State Project Commission. The State Project Commission consisted of representatives from the GKAE, *Minenergo, Gosnadzor*, the USSR Ministry of Health, and the equipment suppliers—the *Minenergomash* and the *Minsredmash*. Significantly, members of the State Project Commission were appointed by *Minenergo*.[21] For each nuclear power station, a separate State Project Commission was formed. After reviewing the project documentation, the State Project Commission issued a signed certificate that formally authorized the operation of the facility as a power producer.[22] This project documentation was not open to public scrutiny and could not be challenged by institutions or individuals outside the State Project Commission.

Decision Criteria and Institutional Bias

The structure of decision-making institutions and the criteria used by those institutions neccessarily influence decision outcomes. In the case of the Soviet nuclear power industry, the relevant institutions and their decision criteria influenced both the spatial and technological character of the industry. Despite some general research done in the area of inter- and intraministry criteria in location and investment decision-making, this subject still remains an area of speculation for Western specialists.[23] Nevertheless, some salient characteristics have been observed in the area of Soviet industrial investment decision-making, including extreme centralization, a strong emphasis on comprehensive planning, and the use of nonprofit decision criteria.[24]

As mentioned before, research and project institutes under the direct subordination of the ministry or as contractual clients carried out project feasibility studies, research and development, regional and branch planning, and engineering cost studies. When evaluating projects, these institutes typically compared different sites, technologies, and engineering parameters, relying largely on cost estimates based on comparisons of analogous facilities built in the past.[25]   At this stage, at the level of the design and research institute, project evaluation was relatively decentralized, with a "chief" or "head" institute involved in overseeing and coordinating other specialized institutes, each responsible for one aspect of the project.[26]   It was the ministry that, after evaluating research findings from the institutes, decided upon a project. The decision was then passed on to high planning agencies, who compared for a given time period the availability of investment resources with requests for investment and future needs (as determined by the party leadership) for additional capacity.[27]

In evaluating alternative projects, cost reduction as a general rule was the fundamental criteria used, at least since the early 1960s. This is not to say that other, noncost-related criteria were not applied, particularly with respect to locational decisions. Indeed Soviet and Western commentators have identified a number of noncost-related "principles" in industrial decision-making, including the promotion of regional economic self-sufficiency, rapid economic development of backward regions, and improvement of the defense capabilities of the country as a whole.[28]   Such "principles" have been formulated from the top down by the Party leadership. However, cost reduction was still the primary criteria upon which these principles were applied.

A number of costing methods were used after the mid-1960s by the Soviets to evaluate projects. In evaluating projects, the methodologies used focused on two factors—minimizing costs and maximizing the returns or effectiveness of capital.[29]   Three cost measures have been identified in evaluating nuclear power projects, these include: specific capital investment (*udel'noye kapitalovlozheniye*), net costs of production (*sebestoimost'*), and the sum of expenses (*privedenniye zatraty*).

According to the 1969 and 1981 Standard Methodology for Determining the Economic Effectiveness of Capital Investments, specific capital investment is defined as

$$K_{sp} = K / C$$

Here specific capital investment K_{sp} equals K the capital investment cost of the plant (in rubles) divided by C, the installed capacity of the plant (measured in kilowatts).[30]   K capital investment includes the design of the station, site preparation, construction of facilities, the cost of equipment and installation, and the cost of start-up and adjustment work. It does not include losses caused by construction (e.g., withdrawal of land previously devoted to agricultural production) and investment in housing.[31]

The net cost of production is defined as

$$S = F + A + L + O$$

Here F is the cost of fuel, A is the charge for amortization of capital, L represents labor costs, and O represents other costs associated with operation.[32]

The sum of expenses is defined as

$$P = S + E_n * K$$

Here P equals S, the net cost of production plus E_n, the coefficient of effectiveness of the investment times K, or capital investment. E_n is in effect an interest charge on capital. Several commentators have noted that the use of E_n was inconsistent in Soviet practice and changed depending on the time it was applied and on the industrial branch being evaluated.[33]

Each of these measures has its shortcomings. The specific capital investment, which is a measure of the relative cost of investment, fails to incorporate important variable costs such as fuel and operating costs. Because nuclear power plants are such capital-intensive projects, and fuel costs are relatively negligible in their cost structure in comparision with fossil fuel plants, this measure is generally unfavorable to nuclear plants unless significant capacities are attained. Conversely, net costs of production, a frequently used measure (usually expressed in kopeks per kilowatt hour), failed to reflect interest charges on capital invested in the plant. Because of the cost structure of Soviet nuclear power plants in which capital costs are high and fuel costs low, this measure showed nuclear projects in a favorable light. The sum of expenses measure is perhaps the most comprehensive method because it reflects an interest charge on capital; however, in practice it was not used consistently and was frequently manipulated to serve the purposes of the evaluator.[34]

The above discussion brings to light several factors that have resulted in an institutional bias toward certain outcomes pertaining to location, scale, and technology in the Soviet nuclear power industry. There are two basic sources of institutional bias in the Soviet electric power industry, these stem from institutional relationships and incentives and from methodological bias in the evaluation process.

The centralization of decision-making in the investment and planning process strongly favored a cost-cutting approach to project formulation. Mindful of obtaining planning authorities' support for the project, the ministry vying for more resources favored projects with lower capital costs and higher capital returns. Thus the relatively narrow and objective studies at the institute level were, when the ministry passed a project proposal to higher authorities, manipulated and pared to generate least-cost approaches. Incentives in the investment and planning process also favored a tendency to lower capital costs. Under the incentive system for project-makers in Soviet industry, which was presumably the same for the nuclear industry, bonuses were awarded for completing of project documentation and construction ahead of schedule and for lowering the investment costs of a planned facility.[35]

The methods used in project evaluation favored certain facility characteristics. In particular, methods used to measure returns from capital favored large-capacity units and facilities. The capital-intensive nature of the nuclear power station cost structure created added pressures for capital cost reduction, particularly through location, and safety, and technology trade-offs.

Siting and Safety Regulatory Framework

As in many other countries, in order to guarantee that the national utility (*Minenergo*) would adhere to safety concerns during the planning, construction, and operation of a nuclear power station, safety regulations and regulatory institutions and procedures were created. To ensure that the site selectors recognized and took into account safety criteria in the process of determining a site, the Soviet state established regulations setting, in most cases, quantifiable limits on various characteristics of the site (see the following chapter). Until 1987 Soviet safety criteria for siting nuclear power facilities covered the following characteristics of the potential site: population density within a specified distance of the proposed facility, meteorological conditions at the site, local geology, proximity to sources of water supply for urban areas, and proximity to hazardous industrial facilities.[36]

As mentioned in the previous section, unlike the decision-making procedures established in several Western countries, there was no formal public licensing process in the Soviet Union. Typically, such a licensing process requires that the utility apply to independent state organizations or local governments for permission to build at the selected site. This way independent institutions are able to validate the safety of the chosen site and project design. Rather, in the case of the Soviet nuclear industry before Chernobyl', the State Project Commission acted as the validator. Nevertheless, beginning with regulations established in 1958, certain state supervisory institutions within the validation process were formally given the power to prevent the construction of a nuclear station at the site selected by *Minenergo*, if the proposed site did not conform to the safety criteria established in siting regulations.[37]

The earliest siting regulations date from December 1958 and remained in effect until 1973. Titled "Provisional Sanitary Regulations for Designing Nuclear Stations," these early regulations restricted nuclear station sites to certain geologic and topographic conditions and placed requirements on population density and the permissible proximity to large urban centers and industrial facilities. Equally important, this document required that the selected site be accepted by the State Supervisor for Sanitation of the USSR Ministry of Health (*Gossannadzor*).[38]

As the scale and scope of application of the Soviet nuclear power industry expanded during the 1970s, a more comprehensive system of normative-technical documents was developed to ensure that safety concerns were addressed during the planning, construction, and operation of nuclear power plants. In the case of siting, the development of safety regulations took on a evolutionary character, which over time, responded to changing safety concerns and to new applications of reactor technology. Three documents came out in the early 1970s, which apparently replaced the "Provisional Sanitary Regulations..." of 1958. These documents, like the 1958 document, placed restrictions on where nuclear facilities could be sited according to population densities in the area surrounding the proposed station, local topography and geology, seismicity, and proximity to other industrial enterprises.[39] These documents were the "Sanitary Norms for the Design of Industrial Enterprises" (1972), the "Basic Sanitary Regulations for Operations with Radioactive Materials and Other Sources of Ionizing Radiation" (1973), and "the General Safety Regulations for Nuclear Stations during Planning, Construction, and Operation" (1973).[40] The latter document—the "General Safety Regulations"—did not set specific standards but rather established general safety guidelines for all areas of the Soviet nuclear industry, to be rigorously quantified in other subordinate documentation.

By the late 1970s, the system of regulatory documentation for nuclear safety gained greater complexity and detail in the form of subsequent subordinate documentation to the "General Safety Regulations..." of 1973. Included in this new set of documentation were further-refined siting regulations, including the "Sanitary Regulations for the Planning and Operation of Nuclear Stations" (1978), the "Requirements for the Location of Nuclear Heat Supply Stations and Nuclear Cogeneration Stations Based on Conditions of Radioactive Safety" (1978), and the "Provisional Norms for the Designing of Nuclear Power Facilities for Seismic Regions" (1979).[41] These documents refined siting requirements for nuclear power stations and modified regulations to apply to new types of power stations envisioned by Soviet planners, in particular, nuclear cogeneration and heat supply stations.

These siting regulations, as well as other nuclear industry safety regulations introduced at that time, were to be enforced by the USSR State Industrial and Mining Technical Inspectorate (*Gosgortekhnadzor*), a new regulatory agency the USSR State Nuclear Inspectorate (*Gosatomnadzor*), and *Gossannadzor*.[42] *Gosatomnadzor* was responsible for monitoring equipment reliability and safety, *Gosgortekhnadzor* for monitoring safety conditions for plant operators as well as operator and reactor management safety, and *Gossannadzor* for monitoring emissions, environmental impacts, and adherence to siting regulations.[43]

By the early 1980s, the Soviets became concerned over a lack of a unified regulatory framework in the nuclear industry. Indeed, several quality control problems were beginning to surface by this time, as the industry rapidly expanded.[44] Additionally, it appears the Soviets became more concerned over safety arrangements in the industry as a result of the U.S. reactor accident at the Three Mile Island plant in 1979.[45] As a result, during the early 1980s, several new organizational changes were made in the regulation and enforcement of safety for the Soviet nuclear power industry. With respect to the siting issue, two salient developments occurred: (1) new siting regulations were established, and (2) regulatory institutions were reorganized.

A new set of general regulations, the "General Safety Regulations for Nuclear Stations during Design, Planning, and Operation" went into effect in 1982.[46] This new document, drawn largely from the "General Safety Regulations" of 1973 and supplementary documentation developed in the later 1970s, placed more stringent siting restrictions on nuclear power stations (hereafter AES).* Additionally, for nuclear heat supply stations (hereafter AST),** more stringent technical safeguards were required of AST reactor designs in view of the planned proximity of ASTs to population centers.

Following the enactment of this new set of regulations, in 1983 there was an organizational streamlining of the institutions previously responsible for safety regulation and monitoring (including siting) in the Soviet nuclear power industry. A new organization, the USSR State Nuclear Power Safety Inspectorate

---

* AES—from the Russian *atomnaya elektricheskaya stantsiya*, atomic power station.
** AST—from the Russian *atomnaya stantsiya teplosnabzheniya*, atomic heat supply station.

(*Gosatomenergonadzor*) was formed. Essentially, *Gosatomenergo-nadzor* represented an amalgamation of *Gosatomnadzor*, *Gosgortekhnadzor*, and *Gossannadzor*.[47] While *Gosatom-energonadzor* was intended to centralize monitoring information and regulatory enforcement, some aspects of the monitoring process, for example, on-site radiation monitoring, were carried out by several other institutions.[48] In addition, *Gosatomenergonadzor* was given increased powers of enforcement relative to previous regulatory institutions.[49] Despite *Gosatomenergonadzor*'s increased powers of enforcement, it was not an independent regulatory agency. Within the policy and planning levels of the Soviet system, it was integrated into the administrative structure of the USSR Council of Ministers responsible for power production.[50]

Characterizing the Institutional Setting

It appears that during the period discussed, the institutional setting for decision-making in the Soviet nuclear power industry conformed strongly to Nelkin and Pollack's notion of an "elitist" framework. Decision-making powers were vested exclusively in the national policy-makers and planners and the branch ministry for electric power production—*Minenergo*. National policy-makers—the upper levels of the party and state—determined policy which was then implemented by *Gosplan* and *Minenergo*. Conversely, these planners, *Minenergo*, as well as several independent research institutes, played a significant role in determining policy in providing information and advice to policy-makers. Within the decision-making process, there were powerful institutional incentives toward cost reduction in construction and operation.

Under the framework existing before Chernobyl', there was no avenue for intervention in site decision-making by local government or local residents. The Soviet framework was intended merely to ensure adequate coordination between project participants and that technically competent decisions were made. Clearly, local concerns and interests were not to be represented, at least formally. Indeed, there is anecdotal evidence that when local officials tried to intervene in nuclear project decision-making, *Minenergo* and *Gosplan* were able to override local concerns. For example, during the 1970s the Ukrainian SSR passed a law requiring a closed-cycle cooling system for all power stations. When local officials attempted to enforce this law during the planning and construction of the South Ukrainian AES, the ministry and *Gosplan* were able to circumvent the law.[51] It was in their interest to do so because the installation of a closed-cycle cooling system would have raised the cost of construction by an estimated 10–12% and lengthened construction time by a year and a half.

It should be stressed, however, that it is unclear to what degree local party officials through the Communist Party apparatus could intervene to influence decision-making. Local party committees at the oblast' (*obkom*), rayon (*raykom*), or city (*gorkom*) levels played an important role in implementing party and state policies. Local party leaders were held responsible for fulfilling key indicators of the plan and were expected to assist and pressure local enterprises into fulfilling targets set from above.[52] Given the priority the party leadership placed on the rapid growth of the nuclear power industry, in all likelihood the professional fate of local party officials would be closely tied to the successful and timely completion of the project. Thus there would be little incentive for local party interference.

Even in the realm of safety and siting regulation, there was little evidence that formal institutions responsible for safety enforcement were able to influence the decision-making process. *Minenergo*'s ability to select the members of the State Project Commission ensured a large degree of control over project validation. Elsewhere in the supervisory and regulatory process, local inspectors were typically employed at the nuclear stations or nuclear projects they were supervising.[53] The power ministry usually prevailed in the ensuing conflict of interest. Indeed, anecdotal evidence in the form of recent revelations about the planning and construction of the Crimean AES indicate that during the initial review of the design draft and the final review by the State Project Commission little attention was paid to siting-related safety concerns or regulations. Project reviewers simply rubber-stamped the project.[54] Similar "rubber-stamping" occurred at the Rostov AES where the construction administration's chief engineer signed the acceptance documents on plant safety in place of the technical inspector.[55] As one Soviet observer has put it: "The government agency which should serve as a guardian of safety in fact was transformed into an agency which legitimizes deviations, i.e., which officially covers up violations of safety requirements."[56]

Thus the decision-making powers in the Soviet nuclear power industry were relegated to the ministry and planning agencies. With little effective oversight from safety enforcement agencies or local institutions, it was the interests of the ministry that predominated. As the next chapter discusses, policies resulting from this institutional setting determined highly controversial aspects of the location, safety, and technology in the Soviet nuclear industry.

Notes

1. Nelkin and Pollak, "Consensus and Conflict Resolution," pp. 65–75; Kitschelt, pp. 67–85.

2. N. P. Fedorenko (ed.), *Programno-tselevoy metod v planirovanii* (Moscow: Nauka, 1982), p. 42; S. L. Prunzer et al., *Organizatsiya, planirovaniye i upravleniye energeticheskim predpriyatem* (Moscow: Visshaya Shkola, 1981), pp. 37–38.

3. Kelley et al., *Energy Research and Development*, p. 11.

4. Hewett, *Reforming the Soviet Economy*, p. 102.

5. S. L. Pruzner, *Ekonomika, oganizatsiya i planirovaniye energeticheskogo proizvodstva* (Moscow: Energoatomizdat', 1984), p. 221.

6. Chung, p.18; CIA, Directorate of Intelligence, *Energy Decision-Making in the Soviet Union: A Reference Aid*, CR-82-12736 (August 1982).

7. Prunzer et al., pp. 37, 40, 41; Chung, p. 18.

8. The influences of regional organizations and institutes in Soviet investment decision-making are documented by Chung, pp. 22–36.

9. Robert W. Campbell, *Soviet Energy Technologies: Planning, Policy, Research and Development* (Bloomington, IN: Indiana University Press, 1980), p. 167.

10. Hewett, *Reforming the Soviet Economy*, pp. 108–114.

11. Marvin Jackson, "Information and Incentives in Planning Soviet Investment Projects," *Soviet Studies*, Vol. 23, No. 1 (July 1971), p. 4.

12. Koryakin, pp. 586–587; E. P. Volkov, "Gosudarstvennyy nauchno-issledovatel'skiy energeticheskiy institut im. G. M. Krzhizhaovskogo," *Elektricheskiye stantsii*, No. 12 (December

1990), pp. 45–48; S. G. Trushin, "Institut Teploelektroproyekt," *Elektricheskiye stantsii*, No. 12 (December 1990), pp. 19–20.

13. David Katsman, "Balance of Plant in Soviet VVER-1000 Reactors: The Case of Side-Mounted Condensers," in Young, p. 36.

14. Minenergo, *Razvitiye elektro-energeticheskogo khozyaystva*, p. 125.

15. Minenergo, *Razvitiye elektroenergetiki soyuznikh respublik* (Moscow: Energoatomizdat, 1988), pp. 9–10.

16. *Izvestiya*, November 27, 1989, p. 2; "Atomnaya Energetika— Nadezhdy vedomstv i trevodi obshchstva," *Novy Mir*, No. 4 (March 1989), pp. 184–193.

17. David Katsman, *Soviet Nuclear Power Plants: Reactor Types, Water and Chemical Control Systems, Turbines* (Falls Church, VA: Delphic Associates, 1986), pp. 53–56.

18. Ibid.; "Institut Teploelektroproyekt," pp. 19–20; A. A. Levental' et al., "Technical Problems of District Heating on the Basis of Nuclear Fuel," *Thermal Engineering*, Vol. 21, No. 11 (November, 1974), pp. 17–22, (English translation from *Teploenergetika*, Vol. 21, No. 11, pp. 10–16); Campbell, p. 166.

19. Katsman, *Soviet Nuclear Power Plants*, pp. 55–56; V. M. Berkovich, "Designing a Nuclear Power Station with 1000 Mw Water-Moderated, Water-Cooled Reactor Units," *Thermal Engineering*, Vol. 21, No. 4 (April 1974), pp. 22–26, (from *Teploenergetika*, Vol. 21, No. 4, April 1974, pp. 18–22); Sergei Voronitsyn, "Concern in Tatar ASSR about Nuclear Power Station to be Built on Kama River," *RFE/RL*, RL 222/83, June 7, 1983, pp. 1–2. Presumably review of the design draft would also include ministries involved in material support and construction of nuclear power plants such as *Gosstroy*, the Ministry of Installation and Construction (*Minmontazh-spetsstroy*) and the Ministry of Power Machine Building (*Minenergomash*). CIA, *Energy Decision-Making*.

20. Katsman, *Soviet Nuclear Power Plants*, pp. 55–56.

21. Joseph Lewin, "The Russian Approach to Nuclear Reactor Safety," *Nuclear Safety*, Vol. 18, No. 4 (July-August 1977), pp. 444–446.

22. Ibid., p. 444.

23. Judith Pallot and Denis Shaw, *Planning in the Soviet Union* (London: Croon Helm, 1984), p. 150; George A. Huzinec, "A Reexamination of Soviet Industrial Location Theory," *The Professional Geographer*, Vol. 29, No. 3 (August 1977), pp. 261–264; for an excellent overview, see Jackson, pp 3–25. Although dated, Jackson provides a comprehensive treatment of the institutions, criteria, and information in the Soviet investment process.

24. Pallot and Shaw, p. 150; Huzinec, pp. 261–264; Jackson, pp. 3–25.

25. Jackson, pp. 4–5; Katsman, "Balance of Plant," pp. 35–36.

26. Jackson, pp. 4–5; Katsman, "Balance of Plant," pp. 35–36.

27. Jackson, p. 27.

28. Huzinec, p. 261.

29. Ibid., p. 262; Kelly et al., "Economics of Nuclear Power," pp. 50–54.

30. Kelly et al., "Economics of Nuclear Power," p. 50.

31. Ibid.

32. Ibid., p. 53.

33. Ibid., p. 54.

34. Ibid.

35. Thorton, p. 153.

36. N. P. Dergachev, "Kriterii vybora ploshchadok dlya issledovatel'skikh tsentov, energeticheskikh ustanovok i ustanovok toplivnogo tsikla v SSSR," in *Siting of Nuclear Facilities* (Vienna:

IAEA, 1975), p. 80; A. M. Petros'yants, *Sovetskoye atomnoye pravo* (Moscow: Nauka, 1986), pp. 115–117.

37. I. R. Stepanov, *Atomnaya teplofikatsiya v rayonakh severa* (Leningrad: Nauka, 1984), pp. 74–75; A. N. Komarovskii, *Design of Nuclear Plants* (Jerusalem: Israel Program for Scientific Translations, 1968), Appendix to Part I (Translated from the Russian source *Stroitel'stvo yadernikh ustanovok*, Moscow: Atomizdat, 1965).

38. Stepanov, pp. 74–75; Komarovskii, Appendix to Part I.

39. Stepanov, pp. 74–75.

40. Respectively, "Osnovniye sanitarniye pravila raboty s radioaktivnymi veshchestvami i drugimi istochnikamy ioniziruyushchykh izlucheniy" (OSP-72), and "Sanitarniye normi proektrirovaniya promyshlennikh predpriyatiy" (SN-245-71), and "Obshchiye polozheniye obespecheniya bezopasnosti atomnykh elektrostantsiy pri proektirovanii, stroitel'stve i ekspluatsii" (OPB-73). Dergachev, pp. 79–84; V. A. Sidorenko et al., "Normirovaniye bezopasnosti atomnykh stantsiy v SSSR" in *Nuclear Power Experience*, Vol. 4 (Vienna: IAEA, 1982), p. 626; P. P. Aleksashin et al., "Razvitiye trebovaniye po bezopasnosti i sistemy gosudarstvennogo nadzora kak osnovy bezopasnogo razvitiya yadernoy energetiki," in *Nuclear Power Performance and Safety*, Vol. 4 (Vienna: IAEA, 1987), p. 428.

41. Respectively, "Sanitarniye pravila proektirovaniya i eksplutatsii atomnykh elektrostantsiy" (SP-AES-78) and "Trebovaniya k razmeshcheniyu atomnykh stantsiy teplosnabzheniya i atomnykh teploelektrotsentraley po usloviyam radiatsionnoy bezopasnosti" (Annex to OPB-73); Sidorenko et al. "Normirovaniye bezopasnosti," pp. 626–627.

42. Ibid., pp. 625–626.

43. Ibid.

44. Gold, p. 20

45. Aleksashin et al., pp. 428–434.

46. Ibid.

47. Ibid.; A. I. Belyaev, "USSR Regulatory and Supervisory Practices in Nuclear Plant Safety," in *Regulatory Practices and Safety Standards for Nuclear Power Plants* (Vienna: IAEA, 1989), pp. 53–54.

48. Actual radiation monitoring is not carried out by *Gosatomnadzor* but by the USSR Ministry of Health, the USSR Ministry of Internal Affiars, and the USSR Hydro-meterorological Committee. Belyaev, pp. 53–54.

49. Ibid., p. 56.

50. A. M. Bukrinskii et al., "Nuclear Plant Safety and Government Regulation," *Soviet Atomic Energy*, Vol. 68. No. 5 (November 1990), p. 382 (from *Atomnaya energiya*, Vol. 68, No. 5, May 1990, pp. 328–332).

51. D. Kolbasov, "Ekologicheskaya politika SSSR," *Sovetsoye gosudarstvo i pravo*, No. 3 (1982), pp. 82–84.

52. Hewett, *Reforming the Soviet Economy*, pp. 106–107.

53. Bukrinskii et al., p. 384.

54. *Izvestiya*, November 27, 1989, p. 2.

55. Thorton, p. 150.

56. Bukrinskii et al., p. 384.

Chapter IV

Policy in the Soviet Nuclear Power Industry, Early Years to 1986

The previous chapter described the formal framework in which Soviet siting decisions were made from the early years of the Soviet nuclear program to 1986. In this chapter, the plans and policies concerning facility location, scale, and technology are examined during the same period, from the mid-1950s to 1986. As pointed out in the last chapter, during this period, site decision-making was dominated by central planners and *Minenergo* officials. The siting policies adopted over this period evolved considerably, reflecting the preferences of the relevant decision-makers and underscoring important changes in the priorities that these decision-makers gave to economy of construction and operation, as well as to safety.

This analysis of Soviet policies is based on the statements of individuals in the decision-making hierarchy, empirical evidence on plant characteristics and technology, and on the changing requirements of siting regulations. This analysis is limited by the fact that not all Soviet regulations on siting are available and by the paucity of data on the site characteristics for each Soviet nuclear facility built, planned, or under construction. Nevertheless, some basic overall trends are discernable for location, scale, and technology of nuclear power facilities in the USSR during the period examined.

Development of the Soviet Nuclear Power Industry

The Soviet nuclear power industry up to 1986 underwent three distinct phases of development. The first, between 1954 and roughly 1970, was characterized by the construction of small experimental and medium-sized prototype industrial reactors intended to explore the economic viability of the commercial use of nuclear energy. These prototype reactors were based on several different reactor designs for a variety of applications. The second stage, beginning in the early 1970s and continuing up to the early 1980s, was characterized by the construction of large-capacity reactors intended for base-load electric power generation at large-capacity facilities in

73

the European USSR. A third stage, planned to begin in the mid- to late 1980s, included the continued rapid expansion of base-load capacity and a diversification of application with the introduction of new types of reactors and facilities for cogeneration (heat and electricity production) and heat production for urban and industrial use. In each of these stages of development, there were changes in siting policy.

The early years of the Soviet nuclear power program focused on the development of experimental and prototype reactors. Between 1954 and the beginning of the 9th Five-Year Plan in 1971, thirteen reactors were built in the Soviet Union and an additional nine reactors had reached an advanced stage of construction (see Appendix A). During this time the reactors brought into operation were small, ranging from 5 to 365 MW in capacity. At least six of these reactors were intended for long-term commercial operation.[*1] Altogether, reactors for commercial power generation had gone into operation at five sites between 1954 and 1970. Construction began at at least four additional sites during this time (See Map 3).[2] Of these nine facilities, eight operated experimental or prototype reactors. The Beloyarsk AES, Bilibino ATETs, Shevchenko ATETs, and Kola AES represented prototype facilities that used further refined reactor designs based on fast-neutron, graphite-moderated water-cooled (RBMK) and pressurized water (VVER) reactor technologies.[3] The Bilibino and Shevchenko ATETs were particularly novel arrangements, intended to cogenerate electricity and hot water.[4] Construction on the Leningrad facility, beginning in March 1970, represented the first truly standardized plant with standardized serially produced reactors for base-load operation.[5] By the end of the 8th Five-Year Plan in 1970, some 897 MW of nuclear power capacity had been installed in the Soviet Union.[6]

Based on the success of several of these prototype nuclear facilities, reactor designers and *Minenergo* officials were able to convince policy-makers and planners of the economic viability of nuclear power as an alternative energy source.[7] Past performance had shown that nuclear power was a cost-effective alternative energy source for power generation in the European Russia, Ukraine, North Caucasus, and Urals and under some conditions in the Far North, if significant economies of scale and standardization of production were realized. Thus, by the 24th Party Congress in April 1971, the decision had been made by policy-makers and planners at the highest levels of the party and state hierarchy to embark on an

---

[*] The Obninsk AES and Siberian AES do not appear to have been in continuous commercial operation.

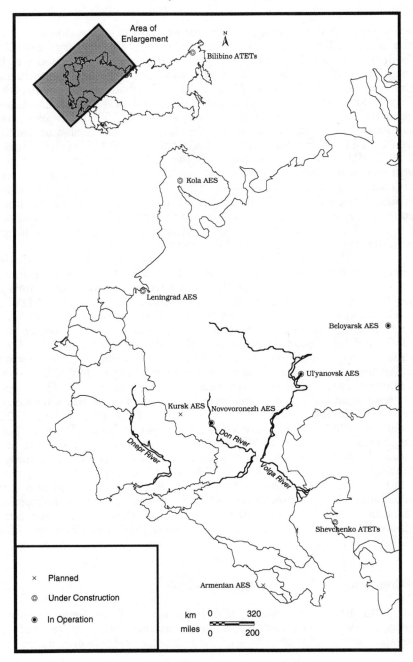

Map 3: Soviet Nuclear Power Stations, 1970

ambitious expansion program for nuclear power generation based on the standardized production of large reactors at large-capacity facilities. At the 24th Party Congress, plans were announced for the installation of 7,000 MW of nuclear-based generation capacity during the ensuing 9th Five-Year Plan (1971–1975).[8] In the following 10th Five-Year Plan (1976–1980), another 13,000 MW of nuclear-based capacity was planned to go into operation.[9] These represented extremely ambitious plans, and ultimately, over the period of the 9th and 10th Five-Year Plans (1971–1980), only 11,700 MW of the planned 20,000 MW went into operation. Between 1971 and 1980, eight new stations went into operation in the USSR and construction had started on at least six (see Map 4). While two of these completed stations, the Bilibino ATETs and the Shevchenko ATETs were small capacity prototype facilities, the remaining six facilities were stations for base-load electric power generation, using medium- to large-capacity reactors (440 to 1,000 MW) of standardized design.

The third stage in the development of the pre-Chernobyl' Soviet nuclear program encompassed the 11th and 12th Five-Year Plans (1981–1990). Soviet plans for this period were extremely ambitious to say the least. For the 11th Five-Year Plan alone, some 25,000 MW of nuclear capacity were planned to be brought into operation.[10] An even greater commitment to the development of nuclear power was confirmed by the announcements of Soviet policy-makers in April 1983 of the "Long-Term Energy Program of 1983." This program called for, among other things, an accelerated expansion of base-load nuclear capacity and the introduction of cogeneration (ATETs) and nuclear heating (AST) stations.[11] On the eve of the Chernobyl' accident, Soviet goals for the 12th Five-Year Plan (1986–1990) included 40,500 MW of capacity to be brought on line.[12] Despite these ambitious plans, only 15,796 MW were installed between 1981 and 1985.[13] Nevertheless, by the end of 1985, nuclear power-based capacity additions comprised 30.5% of total electric power capacity additions over the period 1981–1985.[14] Between 1981 and 1985, an additional six new stations went into operation, with construction starting at another nine sites and no fewer than seven other sites planned for construction start-up between 1986 and 1990 (see Map 5).[15]

Siting, Scale, and Technology, 1971–1980

As mentioned earlier, the early years of the Soviet nuclear program were marked by a very modest program aimed at developing experience with different reactor technologies operating

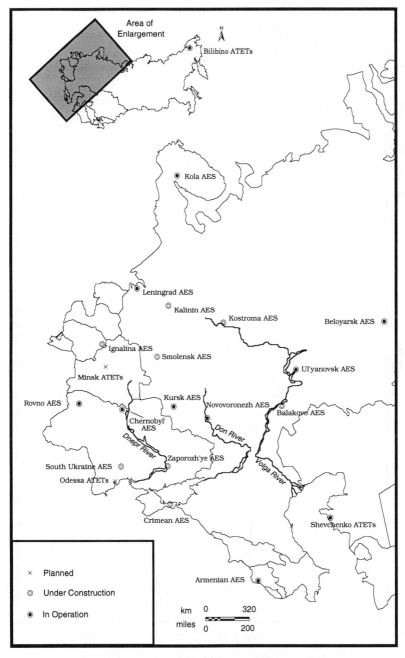

Area of Enlargement

N

Bilibino ATETs

Kola AES

Leningrad AES

Kalinin AES

Kostroma AES

Beloyarsk AES

Ignalina AES

Smolensk AES

Ul'yanovsk AES

Minsk ATETs

Rovno AES

Kursk AES

Novovoronezh AES

Balakovo AES

Chernobyl' AES

Don River

Dnepr River

Zaporozh'ye AES

South Ukraine AES

Odessa ATETs

Volga River

Crimean AES

Shevchenko ATETs

× Planned

◎ Under Construction

◉ In Operation

Armenian AES

km   0   320
miles 0   200

Map 4: Soviet Nuclear Power Stations, 1980

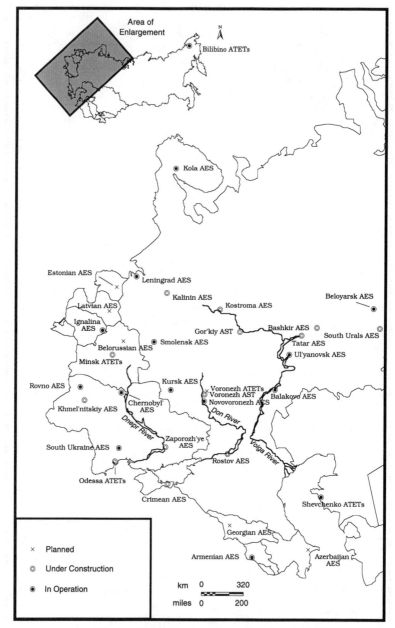

Map 5: Soviet Nuclear Power Stations, 1985

Note: Locations are approximate for Belorussian, Estonian and Latvian AES.

under several different conditions for different applications. During this time, facilities were for the most part located in relatively remote areas, with little concern for cost reduction.

However, the decision to rapidly expand the Soviet commercial nuclear power program around 1970 had major implications for planning, particularly for issues relating to location, plant scale, and technology. At this time the overriding concern of reactor designers and planners was economy of operation and cost reduction. In justifying planners' and policy-makers' investment in nuclear power, reactor designers and power ministry officials argued that with economies of scale and standardized production of existing reactor designs, nuclear power was a competitive source of power in the European third of the USSR and, under some conditions, in remote areas of the Far North. According to the Soviet economists V. V. Batov and Yu. I. Koryakin of the Siberian Institute of Power Engineering of the SO AN SSSR, the *sebestoimost'* (net cost of production) for nuclear-generated power would be competitive with fossil fuel-based power only in the European USSR west of the Urals.[16] Batov and Koryakin stressed that to achieve this competitiveness, nuclear power would have to be produced on a base-load basis with large-capacity units. As a result of independent research, D. G. Zhimerin of ENIN argued that large-capacity nuclear power stations close to load centers in the European USSR could equal or even undercut production costs for fossil fuel electric power generation in this region.[17] Later studies during the 1970s by A. A. Troitsky of the Department of Power and Electrification of *Gosplan* presented similar arguments.[18] Some Soviet researchers also advocated using nuclear power stations in remote areas of the Far North, where the poor infrastructure and the enormous distance from fossil fuel sources made transportation costs for conventional thermal power extremely high.[19]

According to these Soviet researchers, for nuclear power to be a competitive option in the European part of the country, significant economies of scale would have to be achieved over existing designs. As a result, after the late 1960s there was an increasing commitment to reaping economies of scale through larger reactors and larger nuclear power stations. Indeed, early in the nuclear program, Soviet reactor designers and energy economists stressed the cost advantages of large reactors.[20] By 1966, the experience with the graphite-moderated, water-cooled RBMK-type reactors at the Beloyarsk AES and the pressurized-water VVER-type reactors at the Novovoronezh AES confirmed earlier claims that increases in reactor capacity with already established and proven reactor designs would be economical.[21]

A policy commitment to large-capacity reactors was confirmed by the announcement in 1971 that the future expansion of the Soviet nuclear program was to be based on new reactors up to 1,000 MW in capacity.[22] The RBMK design developed by *Gidroproyekt* was selected to be the first of these large reactors. Its modular design allowed it to be assembled on-site and, as a consequence, simplified the design and construction problems associated with the increase in size.[23] The first of the new series of the large-capacity RBMK reactors was the RBMK-1000. This was a standardized design with a gross power generation capacity of 1,000 MW. The first RBMK-1000 was installed at the Leningrad AES and went into operation in 1973. The RBMK-1000 became the mainstay of the Soviet civilian nuclear program during the 1970s and the early 1980s. Indeed, between 1973 and 1985, fourteen RBMKs were installed, amounting to more than 55% of all nuclear capacity additions over that period (see Appendix A).

By the mid-1970s, Soviet reactor designers and power ministry officials advocated even larger reactors based on the established RBMK design. These included the RBMK-1500, a reactor with a gross capacity of 1,500 MW, and the RBMK-2000 with a capacity of 2,000 MW.[24] Construction started on the first RBMK-1500 reactor (at that time the largest reactor in the world) in 1977 at Ignalina; it was subsequently brought on line in December 1983.[25] By 1985, four RBMK-1500 reactors were planned for operation.

A second major reactor design that was adopted for standardized series production was the VVER-type reactor. The VVER-440, a medium-capacity reactor with a gross capacity of 440 MW, was first installed at the Novovoronezh AES in 1971. Between 1971 and 1985, ten VVER-440s went into operation, accounting for approximately 17% of all nuclear capacity additions during that time (see Appendix A). The economic success of the VVER design encouraged the Soviets to adopt an even larger VVER-type reactor, the VVER-1000, during the early 1970s.[26] The VVER-1000 was, like the VVER-440, a standardized design intended for series production with a gross capacity of 1,000 MW. The first VVER-1000 went into operation at the Novovoronezh AES in April 1980 (see Appendix A). The VVER-1000 along with the RBMK-1000 were to be the main reactor types for the Soviet nuclear program throughout the 1980s. Soviet planners envisioned at least ten VVER-1000s to be completed between 1980 and 1985. Nevertheless, between 1970 and 1985, only seven VVER-1000s were brought on line, some 28% of capacity additions for that period.[27]

Thus there was, during the period 1970 to 1985, a rapid shift to large-capacity reactors in the Soviet Union as the nuclear power industry switched from prototype reactors in the range of 100–365 MW to standardized reactors ranging from 440 to 1,500 MW in capacity. The switch to large-capacity reactors is reflected by the fact that between 1965 and 1985 the average capacity per reactor in the USSR increased from 120 MW to 645.3 MW (see Table 12).

In addition to increasing individual reactor capacity, Soviet reactor designers and planners also attempted to gain economies of scale by increasing station or facility size. As in the case with reactor size, this decision seems to have been made around 1970. During the 1960s, Soviet designers and planners envisioned nuclear power stations with capacities of up to 2,000 MW.[28] By the 24th Party Congress in April 1971, Soviet policy-makers indicated plans for very large nuclear power stations with capacities ranging from 6,000 to 8,000 MW.[29] By 1981, according to P. S. Neporozhniy, then minister of *Minenergo*, the majority of nuclear power stations in the Soviet Union would eventually range from 4,000 to 7,000 MW in capacity.[30]

Table 12: Average Reactor Size in the USSR, 1965–1990

	Total gross capacity (MW)	Number of reactors	Mean capacity
1965	360	3	120.0
1970	897	6	149.5
1975	4,767	15	317.8
1980	12,597	27	466.6
1985	28,393	44	645.3
1986	30,393	46	660.7
1990	37,400	45	835.2

Notes: Capacities are at year-end levels. Experimental reactors at the Obninsk AES, Siberian AES, and the Ulyanovsk AES are not included.

Source: See Appendix A.

The first large-capacity stations were planned to be built at the Leningrad AES, the Kursk AES, the Chernobyl' AES, and the Smolensk AES.[31] By 1985, plans included large reactors to be built at least nine other nuclear power stations (see Table 13).[32] The increasing size of individual facilities through the installation of large reactors in larger concentrations over this period is illustrated in Table 14. Between 1965 and 1986, average station capacity

**Table 13: Large-Capacity Nuclear Power Stations and Their
Planned Capacities as of 1985**

Station	Planned capacity (MW)	Completed capacity 1985 (MW)
Kursk AES	6,000	4,000
Smolensk AES	6,000	2,000
Zaporozh'ye AES	6,000	2,000
Ignalina AES	6,000	1,500
Tatar AES	6,000	—
Chernobyl' AES	6,000	4,000
Leningrad AES	4,000	4,000
South Ukraine AES	4,000	2,000
Balakovo AES	4,000	1,000
Khmel'nitskiy AES	4,000	—
Rostov AES	4,000	—
Volgodonsk AES	4,000	—
Rovno AES	3,880	1,818

Note: Selected stations are those that were known to be planned to exceed 3,000 MW in capacity. These amount to more than 70% of planned and existing stations. in 1985.

Source: P. S. Neporozhniy (ed.), *Stroitel'stvo teplovykhi atomnykh elektrostantsiy* (Moscow: Stroyizdat, 1985), pp. 65, 576.

increased from 120 MW to approximately 1,788 MW and the number of reactors per facility increased from 1 per station in 1965 to 2.7 per station by 1986.

The policy of Soviet planners to gain economies of scale through the relative concentration of large standardized reactors at individual facilities placed some serious constraints on plant siting, necessitating their location near large bodies of water. Water consumption rates for large facilities with large reactors are considerable. As early as 1971, G. V. Yermakov claimed that in the process of siting nuclear power stations in the Soviet Union, reliable access to large quantities of water was very critical and was typically a more serious constraint for planners than concerns over radiation safety or transportation costs.[33] According to Yermakov, two VVER-1000 reactors of a total gross capacity of 2,000 MW require 400,000 m^3 of water per hour for coolant purposes, as well as for steam generation.[34] By 1979, Dollezhal' and Koryakin claimed that the lack of water was a serious problem for sites south of Moscow in the European USSR.[35] According to this later source, a nuclear station of 4,000 MW capacity requires a reservoir with a water surface

Table 14: Average Station Size, 1965–1990

	Total capacity (MW)	Number of stations	Mean station capacity (MW)
1965	360	3	120.0
1970	897	3	299.0
1975	4,767	7	681.0
1980	12,597	11	1145.2
1985	28,393	17	1670.2
1986	30,393	17	1787.8
1990 Planned[a]	68,600	23	2982.6
1990	37,585	24	1566.1

Notes: Does not include research facilities.

a Planned refers to planned additions before the Chernobyl' accident.

Source: See Appendix A

of 20–25 km^2.[36]

The increase in reactor size also placed serious infrastructure related constraints on potential sites, particularly those for the large pressurized-water reactors, i.e., the VVER-1000. The VVER design used a cylindrical steel pressure vessel. The reactor and reactor vessel had to be produced off-site, specifically at the Atommash plant in Volgodonsk. These large vessels for the VVER-1000 could only be transported by rail or river, thus further constraining the site availability for this type of reactor.[37]

With the goal of keeping construction and operating costs down, the Soviets, during the 1970s and early 1980s, tended to minimize safety concerns. In developing the RBMK-1000, the RBMK 1500, and the VVER-1000, the Soviets stayed with very simple designs. Moreover, Soviet reactor and facility designs did not possess redundant safety features or safety systems typical of many Western reactors.

Soviet safety philosophy with respect to commercial nuclear reactors was to focus on ensuring the quality of construction of reactor components and to concentrate on sound, simple, and reliable design and operation. Soviet reactor designers did not consider unlikely hypothetical accident events, such as the simultaneous multiple failures of coolant and reactor control systems. According to Soviet scientists and reactor designers, redundant back-up systems, whose need is not evident and reliability questionable, would needlessly complicate reactor operation thus degrading overall safety as well as adding additional costs.[38] As one American observer put it after visiting Soviet reactor facilities in 1964:

> The Soviets do not go as far as does the U.S. in designing for low probability yet credible accidents. They do consider accident potentialities but draw the line at a higher probability level when it comes to adding to plant costs for engineered safeguards or restricting reactor locations because of concern for the major type accident of very low probability.[39]

This philosophy is further evidenced by statements such as the following made in March 1976 by V. A. Sidorenko, then Director of the Kurchatov Institute's reactor program:

> It is not economically justifiable to complicate the design and increase the cost of a power station in order to prevent a shutdown due to a malfunction that is, in practice, almost totally improbable.[40]

Contributing to the Soviet's low concern over multisystem low probability failures was the lack of data on the technical ability to calculate the probability of failure by means of quantitative risk assessment methods akin to those developed in the West. This is not to say, however, that the Soviets did not design without accidents in mind. Soviet reactor designers during this period took into account two types of accidents—internal, or reactor related; and external (due to events external to reactor operation, such as earthquakes or airplane crashes).[41] With respect to internal accidents, reactor safety systems were designed to deal with a variety of potential accidents. For the Soviets, the most extreme credible accident event, the *maksimal'naya proyektnaya avariya* (MPA) or maximum design accident, was a complete rupture of the main coolant pipe of the primary loop.*[42] A large pipe break resulting in the loss of coolant in the primary loop of the reactor was considered the maximum design accident for the RBMK-1000, the VVER-440, and VVER-1000.[43] For accidents of an external nature, the careful selection of the facility site was considered the main safety barrier. For this reason Soviet

---

* The primary loop brings coolant to the immediate vicinity of the reactor core for the purpose of heat removal. Secondary and tertiary loops continue the process of heat removal but limit the spread of radioactivity.

siting requirements, as mentioned earlier, took into account certain types of industrial activity and aircraft overflights.[44]

Although the Soviets used a number of active safety systems for their reactors, which would, under the right conditions, shut down the reactor, several basic design safeguards or passive safety features were ignored to save costs or improve operating performance. One example of this tendency in Soviet reactor design was the positive void coefficient of the RBMK-1000 and RBMK-1500 reactors. This characteristic of the reactor means that as reactor temperature rises, fission activity increases. This is in direct contrast to Western designs, which are characterized by negative void coefficients (i.e., as the reactor temperature rises, fission slows down and finally stops). A positive void coefficient is a highly unstable characteristic and is extremely hazardous under any loss-of-coolant conditions.

Another case of Soviet inadequacies in reactor safeguards involves the containment or localization system for the RBMK and VVER designs. A containment or localization system is intended to provide a barrier to prevent or control the release of fission products into the environment during routine operation and under accident conditions.* Typically, Western reactors use vapor-suppression systems that partially offset the pressure effects of a reactor loss-of-coolant accident, as well as reinforced-concrete structures that are built on the strength and scale to accept the energy, mass, and pressures of a major loss-of-coolant accident. U.S. safety criteria call for such containment to envelop all reactor coolant pipes. [45]

The Soviet approach to containment has been varied, changing over time depending upon reactor type. With the RBMK designs, the Soviets employed perhaps the barest safeguards with respect to localization. The first generation of RBMK-1000s, those preceeding the installation of Chernobyl' 3 and Smolensk 1 (essentially all RBMK-1000s installed before 1980), had no ability to contain radioactive steam leakage in the case of a rupture in the main pipe in the primary coolant system. Although coolant piping was encased in a series of "strong boxes," there was no mechanism to relieve steam pressure from a pipe break via a condensing system.[46] The second generation RBMK-1000s, which began with the Chernobyl'-3 reactor (installed in 1981), were equipped with condensation devices and a pressure suppression pool. With these upgrades, the second-generation RBMK-1000 had the capability of preventing the release of radioactive steam due to a loss-of-coolant accident with a complete

---

* Essentially, containment or localization systems provide for the control of high-temperature and high-pressure gases within the reactor enclosure that result from a rupture in the coolant system of a reactor.

rupture of largest diameter pipe (900 mm) in the primary coolant circuit.[47] Despite these apparent improvements, it is unclear whether these safety features were backfitted to the first-generation RBMK-1000s.

Further illustrating the Soviet policies of economizing on reactor construction and operation through reduced safety systems was the absence of a traditional containment dome. A containment dome is essentially a hermetically sealed shell of steel and prestressed concrete built around the entire reactor and primary coolant system. This type of localization system, also known as a "containment shell," or "secondary containment," is intended as a final safeguard against atmospheric releases of radioactivity, in addition to protecting the reactor from external objects, for example, aircraft or projectiles from a nearby industrial explosion. For the RBMK-series reactors, the Soviets decided against constructing any containment shell. Such containment was considered unnecessary because the nature of the RMBK reactor design precluded the possibility of large-scale releases of radioactive coolant.* Moreover, Soviet reactor designers argued that by avoiding containment shells in their plant designs, several advantages could be gained including: improved accessibility of piping, reactor components, and other machinery, which facilitates maintenance and in some cases reduces thermal stresses on equipment by creating greater space availability for expansion compensation.[48]

With the VVER-440 and VVER-1000 reactors, pressurized-water designs intended for series production, a rather different course toward localization and containment was followed. The VVER-440 and VVER-1000 plant and reactor designs are characterized by greater redundancy in safety control systems (i.e., systems to shut down the reactor in the case of accident or anomalous conditions). The MPA for most VVER-440s and the VVER-1000, like that of the RBMK series, included the complete rupture of the largest diameter pipe (200 mm) in the primary coolant loop. Although it should be pointed out that for first-generation VVER-440s including those reactors built through 1976 (Novovoronezh-3 and -4 and Kola 1 and -2), designers did not envision a large pipe rupture in the primary coolant loop and as a result the units were not equipped with condenser-bubbler equipment.[49] With

---

* This belief was based on the fact that the RBMK reactor consists of more than 1000 individual channels consisting of fuel and surrounding coolant rather than a single vessel encasing fuel and coolant. The compartmentalized nature of the RBMK design thus theoretically precluded the possibility of a sizable release through a rupture of a fuel element or a coolant channel.

respect to localization systems, the Soviets relied on the box or compartmental containment as in the RBMK design, with a series of airtight and pressure-resistant compartments surrounding the pipelines and equipment of the primary coolant loop.[50] In addition, the second-generation VVER-400 and the VVER-1000 were equipped with a system of condenser bubblers as well as sprinklers to condense the steam and thus reduce the pressures created during a loss of coolant accident.[51]

Soviet policy toward secondary containment with VVER-type reactors was somewhat more complicated than in the case of the RBMK. A statement by A. M. Petros'yants, then Chairman of the GKAE, in 1976 indicated that the Soviets did not consider secondary containment shells necessary for any of their reactors and that they did not intend to build them.[52] However, in an article dating from 1974, V. M. Berkovich, a designer with *Teploeletroproyekt*, clearly indicated that the VVER-1000 series reactors would be enclosed by a containment shell of prestressed, reinforced concrete. The containment shell accompanying the VVER-1000 was to be able to withstand the pressures resulting from a rupture of a pipe in the primary coolant loop.[53] Despite the public statements of Soviet officials such as Petros'yants and the opinions of U.S. observers who contend that Soviet plant and reactor designers did not begin to build containment domes until after the American accident at Three Mile Island in 1979,[54] ample evidence suggests that the Soviets were committed to building containment shells for the VVER-1000 reactor. Engineering journal articles by Soviet reactor designers from the late 1960s and 1970s indicate that containment was seriously considered for the VVER-1000 at an early stage of its development. Moreover, a visiting group of U.S. scientists at the Novovoronezh AES in May 1976 noted that construction was under way on a containment shell at the Novovoronezh 5 reactor (the first VVER-1000 to be installed).[55]

Thus, the most likely interpretation of this seeming inconsistency is that Soviet planners and reactor designers since at least the early 1970s, intended to build secondary containment with the VVER-1000. This decision had probably been accompanied by internal debate. However, the completion of the secondary containment system was, in all likelihood, given low priority in the project relative to other goals such as timely reactor start-up. The quick succession of installed containment shells after 1980 may have reflected a higher priority for safety after the Three Mile Island accident.

Siting, Scale, and Technology, 1981–1986

The declining costs at existing nuclear stations during the 1970s and favorable cost estimates for nuclear power in the European USSR convinced policy-makers and planners of the wisdom of an even greater commitment to nuclear power.[56] Thus at the beginning of the 1980s, the Soviet nuclear power industry entered a new stage in its development. This stage was characterized by (1) the enormous scale of expansion of base-load capacity for electric power generation, (2) an emphasis on maintaining and even increasing this rate of expansion, and (3) the diversification of application of nuclear power as an energy source from base-load generation of electric power into heat production for urban and industrial use.

As mentioned before, Soviet plans for the 11th Five-Year Plan called for 25,000 MW of capacity to be installed between 1981 and 1985.[57] On the eve of Chernobyl' in April 1986, Soviet plans envisaged capacity additions amounting to 40,500 MW for the 12th Five-Year Plan (1986–1990).[58] According to P. S. Neporozhniy, Minister of *Minenergo* in 1981, Soviet planners anticipated installing 7,000 MW per year by the end of the 11th Five-Year Plan and 10,000 MW per year by the end of the 12th.[59] By 1990 planners hoped that nuclear power generation would contribute more than one-fifth of all electric power produced in the Soviet Union.[60]

The very scale of the Soviet nuclear program's expansion led to equipment and labor shortages at the reactor sites, causing delays in project completion. Indeed, by 1983 nuclear power plant construction teams at the Tatar AES were advertising for workers as far away as Uzbekistan and Kirgizia.[61] Excessive documentation and bureacratic red tape also seem to have contributed to delays in plant construction. According to one *Minenergo* official, the necessary documentation for site selection as well as technical and economic feasibility studies could take up to seven years to complete.[62] Infrastructure development at the plant site, another prerequisite to plant construction, frequently led to delays amounting to one and a half to two years.[63] In response to the problems being encountered in the Soviet nuclear power industry in meeting planners' goals, a special conference was held by the Central Committee of the CPSU in 1981. This meeting focused on the nuclear power industry's ability to meet the guidelines of the 11th Five-Year Plan.[64]

Even before the 11th Five-Year Plan, concerns were raised during the late 1970s by Soviet plant and reactor designers and economic planners about how best to approach and implement the massive expansion of the Soviet nuclear power industry. One issue that received heightened attention was future Soviet siting policy.

Two Soviet nuclear plant and reactor designers, N. A. Dollezhal' and Yu. I. Koryakin, brought the issue to the forefront between 1976 and 1979 in a series of articles.[65]

According to Dollezhal' and Koryakin, Soviet planners faced an increasingly serious dilemma with the massive expansion of the nuclear power industry. From the planners' and economists' point of view, this dilemma came about from the high desirability, for a large number of large stations and reactors close to population or demand centers in the European USSR. This policy would eventually strain the "ecological capacity" of the densely populated European USSR.* Exceeding the ecological capacity of a given region would, according to Dollezhal' and Koryakin, result in water shortages, land degradation caused by flooding for coolant reservoirs, excessive thermal emissions into water bodies, excessive risk of exposure to populations through waste transport and disposal, and high economic costs associated with expensive safety precautions.[66] To avoid the potential problems that would arise from this dilemma, Dollezhal' and Koryakin advocated building very large agglomerations of stations up to 10,000–50,000 MW in capacity in remote, lightly populated areas with ample water resources. To reduce the risk of contamination of the environment through transport of radioactive waste, waste reprocessing and disposal were to be located at the site of these station agglomerations or "nuclear energy complexes." The authors suggested the Far North of the European USSR as an ideal location.[67]

Dollezhal' and Koryakin's policy proposals were not followed. However, during the early 1980s Soviet planners appear to have greatly expanded the number of large-capacity stations to be built. Indeed Soviet plans on the eve of the Chernobyl' accident called for eleven of twenty-three nuclear power stations to have installed capacities of 3,000 MW or more by the end of 1990 (i.e., the end of the 12th Five-Year Plan).[68] This is in contrast to one station of 3,000 MW or more out of eleven stations in operation at the end of 1980.[69]

However, one element in Dollezhal' and Koryakin's plan does seem to have been adopted in an abbreviated form by Soviet planners. This was the attempt to gain benefits from the agglomerative effects of station location. Specifically, Soviet planners began adopting a new mode of organization for construction work on nuclear power stations. This new mode of organization, the regional construction group or *regional'niy potok*, was first introduced at the Zaporozh'ye

---

* Dollezhal' and Koryakin claimed that according to current plans, some 50 to 70 sites would contain nuclear power stations under construction or in operation by 2000. Dollezhal and Koryakin, "Yadernaya elektroenergetika," p. 27.

and Chigirin AESs during the mid-1980s and was intended to be introduced elsewhere.[70] Under this system, construction groups and equipment would move from one station to the other as each stage of the construction process was completed. Such regional construction groups economized on labor and capital invested in the construction of worker housing and other facilities. The establishment of regional construction groups reflected the tremendous shortages in skilled labor as well as the decreasing distance of Soviet stations from one another as stations became more numerous.

Anecdotal evidence suggests that under the new pressures for massive expansion of the nuclear industry, location decisions were hurried and predicated on a least-cost approach. For example, the Ignalina AES, Rostov AES, and Tatar AES were sited where known geologic conditions or coolant resources were inadequate for the envisioned projects.[71] The experience at the Chigirin AES suggests an extremely cavalier approach to siting.[72] Additionally, the industry expanded into regions with severe energy shortages, for example, the North Caucasus and Caucasus. While the excessive demand for electricity made nuclear stations attractive options for planners, these stations, notably the planned Crimean AES, Krasnodar AES, Georgian AES, and Azerbaijan AES, defied established siting regulations.[73]

During the late 1970s and early 1980s, a new development that had a major impact on Soviet siting and safety policy was the planned introduction of nuclear cogeneration and heat generation stations on a significant scale. Nuclear cogeneration and heat generation stations relied on heat from the reactor to heat water that could then be piped to nearby locations for different applications including residential and municipal heating, as well as industrial uses such as chemical processing and even desalinization.

Soviet interest in harnessing nuclear energy for heat generation dates from the early 1960s. By the mid-1960s, the first nuclear-generated heat for commercial use was produced at the Beloyarsk AES.[74] The heat produced at this station was supplied to station facilities as well as to nearby worker settlements. By the early 1970s, there was considerable support among reactor designers, *Minenergo* officials, and central planners for commercial applications of nuclear-generated heat.[75] At least four basic production modes were envisioned for heat generation using nuclear energy. The earliest developed and most widely adopted method of nuclear heat generation was through the intermittent release or "bleeding" of steam from the turbines at nuclear stations for base-load power generation (AES). A second form involved specialized nuclear cogeneration stations or ATETs equipped with special turbines to cogenerate electricity and heat. Heat production at an ATETs differs from the bleeding of steam at AESs in that the ATETs process uses specialized turbines and produces heat more efficiently. A third mode included nuclear heat stations or AST. ASTs are designed strictly for the production of heat and are even more efficient at this task than the ATETs. Lastly, Soviet planners envisioned very high-temperature nuclear heat stations operating high-temperature gas cooled reactors also known as VTGR (*vysokotemperaturnyy gazookhlazhdayemyy reaktor*—high-temperature, gas-cooled reactor). The VTGR was designed to produce steam at tempertures greater than 200° C for processing industrial products. Whereas, heat produced at the AES, ATETs and AST is intended to supply primarily urban residential and municipal consumers, the VTGR was designed specifically for industrial users.[76]

As mentioned earlier, the first application of nuclear energy for central heating occurred at the Beloyarsk station in 1965 through the bleeding of excess steam from the generating turbines. During the early 1970s, this type of nuclear heat generation had been extended to the Leningrad AES, which was designed to produce up to 70 gigacalories per hour (Gcal/h) per reactor.[77] By 1985 in addition to the Beloyarsk and Leningrad AESs, nuclear heat supply by bleeding excess steam was practiced at at least eight other stations (see Table 15). Heat from these stations was used to supply facilities at the site as well as to nearby workers settlements and towns.[78]

The first nuclear stations designed specifically with heat production in mind were the ATETs cogeneration stations. The first ATETs were the Shevchenko and Bilibino stations, which became operational in 1971 and 1973, respectively. These stations were prototypes intended to provide experience on cogeneration. The Shevchenko ATETs heated water for saltwater desalinization. The Bilibino ATETs generated heat for the centralized heating of residential, municipal, and industrial buildings in a remote mining region in Chukotia in the Far North.[79]

By the mid-1970s, Soviet reactor designers and economists advocated the construction of cogeneration ATETs for large urban centers. Feasibility studies conducted by All-Union Industrial Energy Planning Institute (*Promenergoproyekt*), *Teplo-elektroproyekt*, the Siberian Power Institute of the SO AN SSSR and the High-Temperature Institute of the AN SSSR, indicated that in the European third of the country, ATETs would be economical for large

Table 15: Heat Capacities at Selected Soviet Nuclear Power Stations, 1985

Station	Heat capacity (Gcal/h)
Zaporozh'ye AES	400
Kursk AES	350
Beloyarsk AES	280
Leningrad AES	280
South Ukraine AES	260
Smolensk AES	200
Kola AES	100
Rovno AES	100
Kalinin AES	80
Armenian AES	50

Sources : See Appendix A

cities with heat loads ranging from 1,500 to 3,000 Gkal/h.[*][80] The first large-scale urban ATETs were planned to be operational by the end of the 12th Five-Year Plan (1990). These urban ATETs were to supply large cities with electricity and heated water at a temperature of approximately 170° C for centralized heating. To achieve this economically with existing technology, the maximum distance of transit would have to be less than 20–40 km (12.4–24.8 miles) from the urban area being supplied.[81] Odessa and Minsk were the first cities slated to receive the urban ATETs. Both the Odessa and Minsk ATETs were planned to have two VVER-1000 reactors, each with two TK-500 turbines configured to cogenerate electricity and hot water. The power generation capacity per reactor was 1,000 MW and the heating capacity 900 Gcal/h.[82] With stations this size fulfilling the majority of the power and heating needs of a major urban center rather than a nuclear station and a seperate fossil fuel central-heating station, Soviet planners estimated a savings of some 5 million tons of standard fuel equivalent per year per ATETs.[83] At the beginning of the 12th Five-Year Plan (1986), eight ATETs, each with a total power capacity of 2,000 MW and a heating capacity of 1,800 Gkal/h, were planned for Odessa, Minsk, Voronezh, and Kharkov by 1990; and Gor'kiy (Nizhniy Novgorod), Kuybyshev (Samara), Kiev, and Leningrad by 2000.[84]

---

[*] Heat loads of this range would be typical for a Soviet city of 300,000 to 600,000 people.

The AST concept was also being brought into development during the 1980s. During the 11th Five-Year Plan (1981–1985), construction began on two AST stations at Gor'kiy and Voronezh.[85] Like the ATETs plants, the first units of the Gor'kiy and Voronezh ASTs were to be completed before the end of the 12th Five-Year Plan (1990). The AST configuration included two AST-500 reactors each with a heat-generating capacity of 430 Gcal/h.[86] The AST design was based on the proven pressurized-water reactor concept of the VVER. The AST-500 however included some new, and in the Soviet context, significant improvements in passive safety designs. These design improvements included double-vessel casing around the reactor, reverse pressurization in the coolant loops, and secondary containment.[87]

For the ATETs and AST concepts to be economical, it was necessary to site these stations nearer to urban areas than the AES. New regulations were introduced in 1980 to place new and less restrictive requirements on cogeneration and heat supply stations. According to these siting documents, less restrictive parameters for station proximity were allowed because of safety improvements in reactor and plant design on both the ATETs and AST (see Figure 3)[88]. As Figure 3 illustrates, ATETs were given significantly less restrictions with respect to proximity. The ATETs were allowed within 10 km of cities of 100,000 and within 18 km of cities of 500,000 in population. The new regulations on the AST were however, amazingly unrestrictive, allowing a station to be up to 2 km from a city of any size.[89]

Soviet policies towards location, scale and technology evolved considerably over the period discussed. Until 1970 the Soviet nuclear power industry consisted essentially of a few stations operating small prototype reactors. With the rapid expansion of the industry beginning in 1971, emphasis was placed on large reactors and large stations operating standardized yet simple designs. This emphasis on scale required enormous amounts of water. The Soviet commitment to simplicity in design also resulted in a general relaxation of safety requirements while easing maintenance and presumably construction time. As the Soviet nuclear industry entered the 1980s, even greater emphasis was placed on speeding construction and enlarging plant scale. Additionally, the Soviets planned to introduce cogeneration and heat supply stations. These stations, to save costs, were planned to be located near large urban areas.

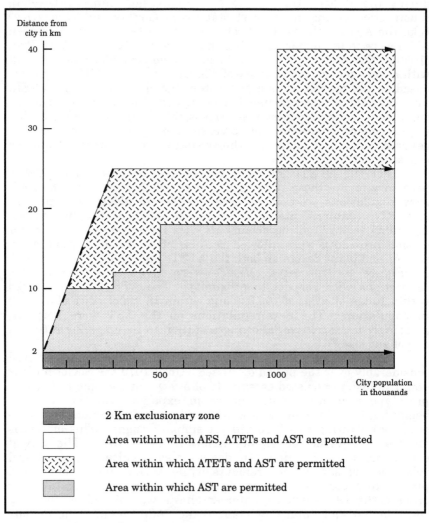

**Figure 3: Regulatory Parameters for Facility Proximity
to Population Centers, 1973-1986**

Source: Sidorenko et al. "Normirovaniye bezopasnosti . . .," pp. 630; "Trebovaniya k razmeshcheniyu. . . ,"
Atomnaya Energiya, pp. 150-151.

Thus, the rapidly expanding industry, in attempting to fulfill the expectations of policy-makers and central planners, adopted designs and projects that increasingly compromised safety and environmental concerns. With little institutionalized outside or independent oversight of designs, industry practices, or project decision-making, the Soviet nuclear industry continued on a road to potential disaster. On April 26, 1986, just such a disaster occurred, calling into question the practices of a previously self-regulated industry.

Notes

1. The Siberian AES operated six 100 MW reactors. Although cited by Soviet sources as a power producer, the Siberian AES was a highly sensitive strategic facility believed to be operated by the Ministry of Medium Machine Building for weapons grade plutonium production. Katsman, *Soviet Nuclear Power Plants*, p. 5; Leslie Dienes and Theodore Shabad, *The Soviet Energy System: Resource Use and Policies* (Washington, D.C.: V. H. Winston & Sons, 1979), 158–159.

2. The five operating nuclear power stations by 1970 included the Obninsk AES, the Siberian AES, the Novovoronezh AES the Ulyanovsk AES, and the Beloyarsk AES. Construction had begun on the Bilibino ATETs, the Shevchenko ATETs, the Kola AES, and the Leningrad AES. IAEA, *Operating Experience with Nuclear Power Stations in Member States in 1987* (Vienna: IAEA, 1988), pp. 473, 498, 513; A. I. Leipunskiy et al., "Construction of an Atomic Electric Power Plant Based on the BN-350 Reactor," *Soviet Atomic Energy*, Vol. 23, No. 5 (November 1967), pp. 1163–1169 (from *Atmonoya Energiya*, Vol. 23, No. 5, May 1967, pp. 409–416).

3. IAEA, *Operating Experience in 1987*, pp. 469, 471, 473; Leipunskiy et al., p. 1163.

4. The combined electric power-generating capacity for the four reactors at the Bilibino ATETs is 48 MW, and the combined heat-generating capacity is 100 Gcal/h. A. M. Petros'yants, *Atomnaya nauka i teknika*, p. 63; IAEA, *Operating Experience in 1987*, p. 471. Depending on the source the rated electric power capacity for the BN-350 reactor at the Shevchenko ATETs is 125,

130, or 150 MW. O. D. Kazachkovskiy et al., "Razvitiye i opit ekspluatsii reaktorov na bistrykh neytronakh v SSSR", *Nuclear Power Experience*, Vol. 5 (Vienna: IAEA, 1982), pp. 21–22; Leipunsky, p. 1163; Lukonin, p. 14. The by-product heated water is used for the desalinization of Caspian Sea water for local consumption (a desalinizing capacity of approximately 120,000 cubic meters of water per day), Leipunskiy, p. 1163; Katsman, *Soviet Nuclear Power Plants*, p. 43.

5. A. M. Petros'yants et al., "The Leningrad Nuclear Power Station and the Outlook for Channel Type BWRs," *Soviet Atomic Energy*, Vol. 31, No. 4 (April 1972), p. 1088 (from *Atomnaya Energiya*, Vol. 31, No. 10, October 1971), p. 317.

6. From data compiled in Appendix A. This figure does not include the Obninsk or Siberian facilities.

7. These officials, researchers, and designers were from energy research and design institutes including: ENIN, Institute of Power Engineering of the Siberian Branch of the Academy of Sciences (SO AN SSSR), and *Minenergo's* power station design institute—*Teploelektroproekt*.

8. "The Directive for the Five-Year Plan-Resolution of the 24th Party Congress of the CPSU Central Committee's Draft Directives of the 24th CPSU Congress for the Five-Year Plan for the Development of the National Economy in 1971–1975," *CDSP*, Vol. 23, No. 18 (June 1, 1971), p. 13. (from *Pravda*, April 11, 1971, p. 1.); V. A. Ryl'skiy, *Elektroenergeticheskaya baza economicheskykh rayonov SSSR* (Moscow: Nauka, 1974), p. 31.

9. "Guidelines for the 10th Five-Year Plan-II," *CDSP*, Vol. 28, No. 16 (May 16, 1976), pp. 2–8 (from *Izvestiya*, March 7, 1976, pp. 2–8).

10. "Guidelines for the 11th Five Year Plan," *CDSP*, Vol. 33, No. 16, (May 20, 1981), p. 11 (from *Izvestiya*, March 5, 1981, p. 3).

11. N. M. Sinev and B. B. Baturov, *Ekonomika atomnoy energetiki* (Moscow: Energoatomizdat, 1984), pp. 60–61.

12. A. A. Troitsky (ed.), *Energetika SSSR, 1986–1990 godakh* (Moscow: Energoatomizdat, 1987), p. 173.

13. Compiled from data in Appendix A.

14. Goskomstat, *Kapital'noye stroitel'stvo*, p. 53.

15. *Operating Experience in 1987*, pp. 491, 465, 487, 535, 539, 543; A. M. Petros'yants, *Atomnaya nauka i teknika* (Moscow: Energoatomizdat, 1987), pp. 22, 24.

16. Kelley et al., pp. 58–59; Yu. I. Koryakin, "Mathematical Modeling of the Developing Nuclear Power Generation," *Soviet Atomic Energy*, Vol. 36, No. 6 (December 1974), pp. 586–591 (from *Atomnaya Energiya*, Vol. 36, No. 6, June 1974, pp. 419–422).

17. D. G. Zhimerin, "The Present and Future of the Soviet Power Industry," *Thermal Power Engineering*, Vol. 17, No. 3 (March 1970), pp. 5–6 (from *Teploenergetika*, Vol. 17, No. 3, March 1970, p. 4).

18. A. A. Troitskiy, "Elektroenergetika: Problemi i perspektivi," *Planovanye khozyaistvo*, No. 2 (1979), pp. 20–22.

19. A. P. Aleksandrov, "Nuclear Power Problems," *Soviet Atomic Energy*, Vol. 13, No. 2 (March 1963), p. 710, (from *Atomnaya Energiya*, Vol. 13, No. 2, August 1962, pp. 109–124).

20. A. M. Petros'yants, "A Decade of Nuclear Power Engineering," *Soviet Atomic Energy*, Vol. 16, No. 6 (June 1964), pp. 596–601 (from *Atomnaya Energiya*, Vol. 16, No. 6, June 1964, pp. 479–484); Zhimerin, p. 5.

21. A. S. Gorshkov, "Soviet Heat and Power Engineering on the Eve of the 23rd CPSU Congress," *Thermal Power Engineering*, Vol. 13, No. 3 (March 1966), pp. 4–5 (from *Teploenergetika*, Vol. 13, No. 3, March 1966, pp. 2–4).

22. "The Directives for the 9th Five-Year Plan," *CDSP*, Vol. 23, No. 18 (June 1, 1971) p. 13 (from *Pravda*, April 11, 1971, p.1).

23. N. A. Dollezhal' and Yu. I. Koryakin, "Yadernaya elektro-energetika: dostizheniye i problemy," *Kommunist*, No. 14 (September 1979), p. 21.

98 Chapter IV

24. N. A. Dollezhal' and I. Ya. Emel'yanov, "Experience in the Construction of Large Power Reactors in the USSR," *Soviet Atomic Energy*, Vol. 40, No. 2 (August 1976), pp. 144–145 (from *Atomnaya Energiya*, Vol. 40, No. 2, February 1976, pp. 117–126).

25. IAEA, *Operating Experience in 1987*, pp. 487.

26. V. M. Berkovich et al., "Designing a Nuclear Power Station with 1,000 MW Water-Moderated, Water-Cooled Reactor Units," *Thermal Engineering*, Vol. 21, No. 4 (April 1974), pp. 22–23, (from *Teploenergetika*, Vol. 21, No. 4, April 1974, pp. 18–22).

27. Kelly et al., "The Economics of Nuclear Power," p. 50.

28. G. V. Yermakov, "Nuclear Power is the Basic Trend in the Development of Future Power Engineering," *Thermal Power Engineering*, Vol. 18, No. 4 (April 1971), pp. 10–16 (from *Teploenergetika*, Vol. 18, No. 4, April 1971, pp.6–11).

29. "The Directive for the 9th Five-Year Plan," p. 13.

30. "Soviet Nuclear Power Plans Outlined," *CDSP*, Vol. 33, No. 22 (June 21, 1981), p. 6 (from *Pravda*, June 4, 1981, p. 2).

31. Petros'yants et al. "The Leningrad Nuclear Power Station," p. 1088; and Appendix A.

32. P. S. Neporozhniy (ed.), *Stroitel'stvo teplovykhi atomnykh elektrostantsiy* (Moscow: Stroyizdat, 1985), pp. 65, 576.

33. Yermakov, p. 16.

34. Ibid.

35. Dollezhal' and Koryakin, "Yadernaya elektroenergetika," pp. 20–22.

36. Water losses through evaporation from such reservoirs amounted to 2 cubic kilometers in 1979. Ibid.

37. Kelly et al., "The Economics of Nuclear Power," p. 50.

38. Gold, p. 33; Lewin, pp. 442–443.

39. Lewin, p. 440.

40. V. A. Siderenko, "Present-day Problems of Safe Operation of Nuclear Power Stations," *Thermal Power Engineering*, Vol. 33 No. 3 (March 1976), p. 9 (from *Teploenergetika*, Vol. 33, No. 3, March 1976, p. 4).

41. V. A. Siderenko, et al., "Razvitiye podkhoda k resheniyu voprosov bezopasnosti atomnykh istochikov energosnobzheniya v SSSR v svyazi s rasshireniyem masshtaba i oblasti ikh primeneniya," IAEA, *Current Nuclear Power Plant Safety Issues*, Vol. I (Vienna: IAEA, 1981), p. 266.

42. P. P. Aleksashin et al., pp. 428–429.

43. This early model of the RBMK-1000, which is identified as those that were built before the Chernobyl' 3 unit in 1981 (construction began in 1977), did not have the localization systems to contain radioactivity releases from the MPA involving loss of coolant in the primary loop. M. Donahue et al., "Assessment of the Chernobyl'-4 Accident Localization System," *Nuclear Safety*, Vol. 28, No. 3 (July–September 1987), pp. 298–299; V. A. Sidorenko et al., "Safety of VVER Reactors," *Soviet Atomic Energy*, Vol. 43, No. 12, (May 1978) p. 1104, (from *Atomnoya Energiya*, Vol. 43, No. 12, December 1977, pp. 449–457).

44. Sidorenko et al., "Razvitiye podkhoda k resheniyu," p. 266.

45. Donahue et al., pp. 306–307.

46. Ibid., p. 298.

47. Ibid., pp. 298-300.

48. Lewin, p. 443; E. P. Anan'yev and G. N. Kruzhilin, "Radioactive Safety Barriers in Nuclear Power Stations," *Soviet Atomic Energy*, Vol. 37, No. 1 (July 1974), pp. 699–705 (from *Atomnaya Energiya*, Vol. 37, No. 1, January 1974, pp. 369–373.

49. Sidorenko et al., "Safety of VVER Reactors," pp. 1103–1104.

50. Ibid., pp. 1100–1101, 1105.

51. Ibid., pp. 1105–1106.

52. Lewin, p. 442.

53. V. M. Berkovich et al., pp. 26–28.

54. David Marples, *Chernobyl' and Nuclear Power in the USSR* (New York, NY: St. Martin's Press, 1986), p. 110.

55. Lewin, p. 442.

56. Kelley et al., "The Economics of Nuclear Power," p. 55.

57. "Soviet Nuclear Power," *CDSP*, pp. 5–6. (from *Pravda*, June 4, 1981, p. 2).

58. Troitskiy, *Energetika v SSSR, 1986–1990 godakh*, p. 173.

59. "Soviet Nuclear Power," pp. 5–6 (from *Pravda*, June 4, 1981, p. 2).

60. Troitskiy, *Energetika*, p. 174.

61. Voronitsyn, "Concern in Tatar ASSR," p. 2.

62. "Leaders Address 27th Party Congress—VII," *CDSP*, Vol. 38, No. 15 (May 14, 1986), p. 10 (from *Pravda*, March 5, 1986, p. 3).

63. Dollezhal' and Koryakin, "Yadernaya elektroenergetika," pp. 20–22.

64. This meeting focused on accelerating the expansion of nuclear capacity. Safety issues were conspicuously absent in the discussions. Gold, p. 20; "Conference in the CPSU Central Committee," *CDSP*, Vol. 33, No. 33 (September 16, 1981), p. 19 (from *Pravda*, July 16, 1981, p. 2).

65. "Atomic Reactor: Light and Heat," *CDSP*, Vol. 27, No. 28 (August 11, 1976), pp. 18–19 (from *Pravda*, July 14, 1976, p. 6) N. A. Dollezhal' and Yu. I. Koryakin, "Nuclear Energy Complexes and the Economic and Ecological Problems of Nuclear Power Development," *Soviet Atomic Energy*, Vol. 43, No. 5 (November 1977), pp. 1019–1024 (from *Atomnaya Energiya*, Vol. 43, No.5, May 1977, pp. 369–373).

66. Dollezhal' and Koryakin, "Yadernaya elektroenergetika," pp. 27, and Dollezhal and Koryakin, "Nuclear Energy Complexes," pp. 1021, 1024.

67. Dollezhal and Koryakin, "Nuclear Energy Complexes," pp. 1021, 1023–1024.

68. Troitskiy, *Energetika*, p. 175.

69. See Appendix A.

70. Petros'yants, *Atomnaya nauka*, p. 48.

71. Saulius Girnius, "Continued Controversy over Third Reactor at Ignalina Atomic Power Plant," Radio Liberty—Baltic Area, SR 18 (August 4, 1988), p. 30; "Rally Protests Tatar AES Construction," *FBIS*, FBIS-SOV-90-123, June 26, 1990, p. 105; *Izvestiya*, August 11, 1990, p. 2.

72. Roman Solchanyk, "Ukrainian Writers Protest against Nuclear Construction Site," *RLR*, RL 336/87, August 11, 1987, pp. 1–2.

73. "In Armenia, Elsewhere," FBIS, FBIS-SOV-88-248 (December 27, 1988), p. 67.

74. Petros'yants, *Atomnaya nauka*, p. 48.

75. V. P. Koryanikov, "Present-Day Conditions and Prospects for the Development of District Heating," *Thermal Power Engineering*, Vol. 19, No. 4 (April 1972), pp. 1–6 (from *Teploenergetika*, Vol. 19, No. 4, April 1972, pp. 2–5); "Atomic Power Plants Under Way: No Hazards Seen," *CDSP*, Vol.. 23, No. 5 (March 2, 1971), p. 25 (from *Ogonyok*, No. 51, December 1970, pp. 6–7; A. M. Petros'yants et al., "Prospects of the Development of Nuclear Power in the USSR," *Soviet Atomic Energy*, Vol. 31, No. 4 (April, 1972), pp. 1067-1074 (from *Atomnaya Energiya*, Vol. 31, No. 4, October 1971, pp. 315–323).

76. Petros'yants, *Atomnaya nauka*, pp. 33, 62.

77. Petros'yants et al. "The Leningrad Nuclear Power Station," p. 1088

78. Several stations supply heat to sizable cities in the USSR. The
    South Ukraine AES supplies Konstantinovka, the Kalinin AES
    supplies Udomlya, the Zaparozh'ye AES supplies Energodar, the
    Rovno AES supplies Kuznetsnovsk, and the Beloyarsk AES
    supplies nearby Zarechniy. Petros'yants, *Atomnaya nauka*, p.
    64.

79. Petros'yants et al., "Prospects of the Development," pp. 1068–
    1069; V. M. Abramov, "The Bilibino Nuclear Power Station,"
    Soviet Atomic Energy, Vol. 35, No. 5 (May 1974), pp. 977–978,
    (from *Atomnaya Energiya*, Vol. 35, No. 5, November 1973, pp.
    299–304).

80. Korytnikov, pp. 1-6; A. A. Il'kevich et al., "Investigating the
    Optimum Unit Capacity and the Composition of the Main Items
    of Plant for Nuclear Heat and Power Stations," *Thermal Power
    Engineering*, Vol. 21, No. 2, (February, 1974), pp. 4–5 (from
    Teploenergetika, Vol. 21, No. 2 February 1974, pp. 3-7); G. P.
    Levental' et al., p. 17.

81. Ya. A. Kovylyanskii, "Centralized Heat Supply from Nuclear
    Sources," *Thermal Power Engineering*, Vol. 28, No. 3 (March
    1981), pp. 132, (from *Teploenergetika*, Vol. 28, No. 3, March 1981,
    pp. 10–16); Yu. I. Tokarev (ed.), *Yaderniye energeticheskiye
    ustanovki* (Moscow: Energoatomizdat, 1986), p. 201.

82. Ibid.

83. A. P. Shadrin, *Atomniye elektrostantsii na kraynem severe*
    (Yakutsk: Yakutskiy filial SO AN SSSR, 1983), pp. 28–29.

84. Kovylyanskii, p. 132; Matthew Sagers, "News Notes," *Soviet
    Geography*, Vol. 27, No. 4 (April 1986), p. 261.

85. "Soviet Nuclear Power Plans Outlined," *CDSP*, Vol. 33, No. 22
    (June 21, 1981), pp. 5–6 (from *Pravda*, June 4, 1981, p. 2).

86. I. I. Nigmatulin and B. I. Nigmatulin, *Yaderniye ener-
    geticheskiye ustanovki* (Moscow: Energoatomizdat, 1986), p. 137.

87. Yu. G. Nikiporets et al., "Bezopasnost' atomnykh stantsiy
    teplosnabzheniya v SSSR," *Nuclear Power Experience*, Vol. 4
    (Vienna: IAEA, 1983), pp. 150–152; A. P. Aleksandrov,

Yadernaya energetika, chelovek i okruzhayushchaya sreda (Moscow: Energoatomizdat, 1984), pp. 72–73, 162–163.

88. Siderenko et al., "Normirovaniye bezopasnosti," p. 630.

89. Ibid.; "Trebovaniya k razmeshcheniyu atomnykh stanstiy teploshabzheniya i atomnykh teploelektrotsentraley po usloviyam radiatsionnoy bezopasnosti," *Atomnaya energiya*, Vol. 49, No. 2 (August 1980), pp. 150-151.

Chapter V

Post-Chernobyl' Changes in Nuclear Policy

In the wake of the Chernobyl' accident,* the Soviet nuclear power industry came under heavy criticism from international and Soviet experts, as well as the international and domestic public. In response, Soviet policy-makers and the Soviet nuclear industry undertook measures to improve safety within the industry and to restore public confidence in the nuclear industry. Soviet policy initiatives or responses affected the siting issue in a number of ways, both directly and indirectly. These policy changes related to the industry's organization and management, siting regulations and criteria, reactor technology, and information policy. Significantly however, until 1990 there was no attempt to include other elements of Soviet society in the decision-making process. Indeed, as this chapter illustrates, Soviet energy decision-makers struggled bitterly to retain absolute control over the decision-making process, particularly at the ministerial level.

Institutions and Management

Initially, Soviet responses to the accident were disciplinary in nature with widespread personnel changes at the top levels of management in the nuclear power industry. These personnel

---

* On April 26, 1986 the Chernobyl'-4 reactor exploded resulting in the largest nuclear accident in history. The accident involved a second-generation RBMK-1000 at the Chernobyl' AES in the north central Ukrainian SSR. The accident was the result of an experiment by reactor operators that went awry. During a routine reactor shutdown, operators wanted to measure how long generating turbines could continue to generate power through mechanical inertia under conditions of a power failure or emergency steam cutoff. To conduct the test, automatic reactor safety systems were switched off and many control rods were withdrawn. As power levels were reduced during the test, cooling water pumps failed and power levels surged. As a result, the fuel rods melted reacting with vaporized cooling water in the reactor causing a buildup in hydrogen and, within seconds, a violent explosion. The reactor was totally destroyed, the building breached, and radioactive particles and gases released into the atmosphere. (M. Ishikawa, "An Examination of the Accident Scenario at the Chernobyl' Nuclear Power Station," *Nuclear Safety*, Vol. 28, No. 4 (October–December 1987), pp. 449–450.

changes occurred in July 1986 and included the replacement of the head of the *Gosatomenergonadzor*; the First Deputy Minister of *Minenergo* (this position was responsible for nuclear stations operated by the ministry); the First Deputy Minister for Medium Machine Building (the ministry specifically responsible for reactor construction); a prominent electric power engineer and corresponding member of the USSR Academy of Sciences; and a deputy director of a research and design institute.[1] Other personnel changes and party explusions (an act that before 1989 represented a severe blow to the career of a industrial administrator) followed within several months of the accident.[2]

More significantly however, there were changes in the administrative structure of the Soviet nuclear power industry. In July 1986 an entirely new ministry—the Ministry of Nuclear Power (*Minatomenergo*) was created from the branch of *Minenergo* responsible for the operation of nuclear power stations. Such a ministerial reorganization as this was entirely consistent with past party leadership experience when faced with a serious and highly visible setback in industry or agriculture. Under this new scheme, the new ministry *Minatomenergo* was to administer the construction and operation of the nuclear power stations formerly under *Minenergo*.[3] Theoretically, this reorganization would focus responsibility and simplify the administration of nuclear power stations.

During the summer of 1989, the administrative structure of the Soviet nuclear power industry was again reorganized. In July 1989 a new ministry—the Ministry of Nuclear Power and Nuclear Industry (*Minatomenergoprom*)—was created. This new ministry represented an amalgamation of *Minatomenergo* and those administrative divisions within the *Minsredmash* responsible for nuclear fuel processing, the production of nuclear reactors and other plant equipment.[4] In addition, *Minatomenergoprom* was given responsibility for nuclear-related defense equipment.[5] The purpose behind this new administrative move was to integrate vertically those organizations and enterprises responsible for the production of equipment, construction, maintenance, and operation of nuclear power stations under one ministry. Other changes followed in the area of operational inspection and reactor management. Perhaps the most significant organizational change from the standpoint of safety was the formation of a new safety supervisory committee to replace *Gosatomenergonadzor*. This new committee, the State Nuclear Industry Supervisory Committee (*Gospromatomnadzor*) while fulfilling the same function as *Gosatomenergonadzor*, was given greater administrative independence from the ministries responsible for power production.[6]

Reactor Technology

The Chernobyl' accident involving a RBMK-1000 reactor forced the Soviet power engineers to reevaluate their reactor designs. Almost immediately Soviet designers took steps to solve the apparent flaws of the RBMK design, which had contributed to the accident. Over the longer term, designers and policy-makers made several important policy decisions about reactor technology.

Within a year of the Chernobyl' accident, a number of changes were made to the RBMK design. Reactor control rods* were engineered to prevent their complete withdrawal from the reactor core.[7] Additionally, the number of control rods was increased in the RBMK-1000 and RBMK-1500 reactors. An improved, automated control system was also installed that reduced the time involved in automatic shutdown from 18–20 seconds to 10–12 seconds.[8] Soviet engineers also introduced design changes aimed at reversing the positive void coefficient of the RBMK design. This included reconfiguring the shape of the graphite blocks surrounding the fuel channels and introducing a higher enriched fuel (from 2.0 to 2.4%).[9]

Despite these safety modifications to the RBMK design, criticism of this design continued both in the USSR and abroad.[10] As early as May 1987, *Minatomenergo* officials stated publically that the RBMK was to be discontinued in any future projects.[11] Despite these statements, existing Soviet plans did appear to include the ten RBMK-1000 reactors then under construction. During 1988 a debate arose among Soviet reactor designers and nuclear power ministry officials over the possibility of continuing the RBMK design, albeit in a greatly modified form.[12] However, a definite decision was made against the continuation of the RBMK design in April 1989, when the USSR Council of Ministers announced that there would be no more RBMK reactors brought into operation at existing nuclear stations. This decision included the Chernobyl'-5 and -6 reactors, the Smolensk-4, and Kursk-6 reactors, all of which were nearing completion.[13] This decision affected a total of four stations and resulted in the cancellation of 5,500 MW of planned capacity.[14]

---

* Control rods are movable rods that absorb neutrons in the reactor core. Through the insertion or removal of control rods, the reactivity in the reactor can be controlled.

The Chernobyl' accident forced the Soviet nuclear power industry to reassess other elements of their reactor technology. The VVER design represented the other major reactor type used by the Soviet nuclear industry, accounting for approximately 41% of the USSR's installed capacity at the time of the Chernobyl' accident.[15] Even before the accident, nuclear industry plans envisioned an even greater role for the VVER design with some 70% of capacity additions to be from the VVER-1000.[16] The VVER's future role was even more prominent as it gradually became apparent that the RBMK had fallen out of favor among nuclear industry decision-makers.

Soviet energy officials placed considerable emphasis on improving the safety of the VVER-1000 reactor after the Chernobyl' accident. This was in part due to the general increase of safety concerns as a result of the Chernobyl' accident. Additionally however, this was a result of the long-standing safety concerns about the pressurized-water design of the VVER.[17] The Chernobyl' accident validated the concerns of those nuclear power engineers who advocated safer pressurized-water designs. As a result, it was announced in October 1988 that two new VVER-type reactors based on new and innovative passive safety concepts would be introduced in the 1990s.[18] The two new designs, designated the VVER-88 (to be introduced in 1993) and the VVER-92 (to be introduced in 1999), incorporated passive safety features such as a passive decay heat removal system, a passive system for injection of boron into the reactor to ensure reactor shutdown in the event of a failure of the automated systems, and a series of design features for immobilizing and cooling the core under conditions of a core meltdown.[19]

Many of these safety features were first developed for the AST-500. Their adoption to the VVER-1000 design represented a fundamental shift toward a uniform level of safety among nuclear heating stations and conventional base-load power stations. Significantly, these passive features took into account accident possibilities that before Chernobyl' were not considered credible accident events. According to Soviet experts, the MPA for both the RBMK and VVER designs had been redefined. With the VVER-88 and VVER-92, Soviet designers were attempting to exclude the possibility of core meltdown or other catastrophic accident situations with a probability of 10^{-5} to 10^{-6} per reactor year.[20] This was in contrast to past policy which was concerned with accident events with a probability of 10^{-3} to 10^{-4} per reactor year.[21] Further illustrating greater Soviet concern over low-probability accidents, concerted efforts were launched to incorporate Western-style probabilistic risk assessment (PRA) procedures into reactor design and assessment.[22]

Another area in which Soviet policy toward reactor technology changed was the ATETs program. In 1986 four sites for ATETs at an advanced stage of planning or construction existed at Odessa, Minsk, Khar'kov, and Volgograd. These ATETs were to be operational by the early 1990s with another four stations to be built at Gor'kiy (Nizhniy Novgorod), Kuybyshev (Samara), Kiev, and Leningrad by 2000. By 1988 preliminary construction began for another ATETs at Yaroslavl' in Belorussia. The ATETs design, as mentioned in Chapter IV, was based on the VVER-1000 design with specialized turbines for cogeneration of heat and electricity. In contrast to the AST-500, the ATETs scheme had few safety additions to the basic VVER design, although the requirements for economic heat distribution neccessitated its location within 40 km of consumer centers (i.e., large urbans areas).[23] In September 1988, Alexander Protsenko, chairman of the GKAE, announced that the ATETs projects for Minsk and Odessa were permanently canceled and those at Khar'kov and Volgograd were postponed until a new, safer, and more economical ATETs design became available.[*24]

Soviet reactor policy also acquired a new element after Chernobyl', the adoption of a new and safer line of reactor designs for cogeneration and heat production—high-temperature, gas-cooled reactors (VTGR). Before Chernobyl', considerable research had been undertaken in VTGR reactor technology, and the design had been advocated by several Soviet reactor designers.[25] Nevertheless, the VTGR represented a completely new line of thinking in Soviet commercial reactor technology. The design did not possess the extensive development background of the RBMK, VVER, or even fast-breeder designs used for commercial power generation. Moreover, Soviet VTGR designs envisioned small-capacity reactors (i.e., less than 500 MW capacity), a notable departure from the trend toward large-capacity reactors. The VTGR design, while forfeiting economies of scale, offered several advantages including inherent safety and high-temperature steam available for a wide array of applications.

Less than a year after the Chernobyl' accident, in January 1987, energy policy-makers from the GKAE, *Gosplan* , and the AN SSSR recommended a series of policy initiatives in the party's Program for Scientific and Technical Progress for 1991–2010. For the Soviet nuclear program, two developments were explicitly outlined.

---

* Presumably the Yaroslavl ATETs was not mentioned because this project was at a very early stage and would not be affected by the delay in reactor reevaluation and redesign.

First was the emphasis on new and improved VVER-1000 designs. Second was the announcement that two nuclear cogeneration designs, the VTGR-250 and the VG-400, were to be developed and brought into operation.[26] There was considerable Soviet interest in Western, and in particular West German, gas-cooled reactor designs. This interest culminated in an agreement signed on October 24, 1988, between a Soviet design enterprise under the direction of GKAE and the German firms Kraftwork Union* and Asea Brown Boveri AG to build jointly a 200 MW high-temperature, gas-cooled reactor.[27]

Siting Policy

Despite policy changes in issue areas that directly contributed to the Chernobyl' accident—reactor management and design—Soviet energy policy-makers, nuclear industry specialists, outside experts, as well as the general public, all argued that a reassessment of nuclear station siting policy was in order after Chernobyl'.[28] In contrast to management and technology policy reversals, Soviet officials were surprisingly reluctant to discuss in detail new siting policies.

Soon after the Chernobyl' accident in April 1986, Soviet officials indicated that Soviet siting procedures and criteria were under reevaluation.[29] In October 1987, a new and apparently comprehensive siting document went into effect.[30] Not all provisions of the new document have been disclosed to the public. Indeed, only provisions encompassing proximity to populated areas, plant scale, and plant type have been described openly. Comments by N. F. Lukonin, then Minister of *Minatomenergo,* indicate that proximity to populated areas, seismicity, local geology, and local meterology were included in the new post-Chernobyl' siting regulations.[31]

The October 1987 siting regulations were based on a different set of reactor analyses than that previously done. The new siting criteria took into account accident situations and consequences not evaluated in the past analyses used for the formulation of siting regulations OPB-82 and SPAES-79.[32] The most significant change in the new siting regulations involved restrictions on plant proximity to urban areas. The October 1987 regulations prohibit the location of an AES closer than 25 km from cities of more than 100,000 inhabitants (see Figure 4).[33] This was in stark contrast to earlier Soviet siting

---

* KWU is a division of Seimens AG.

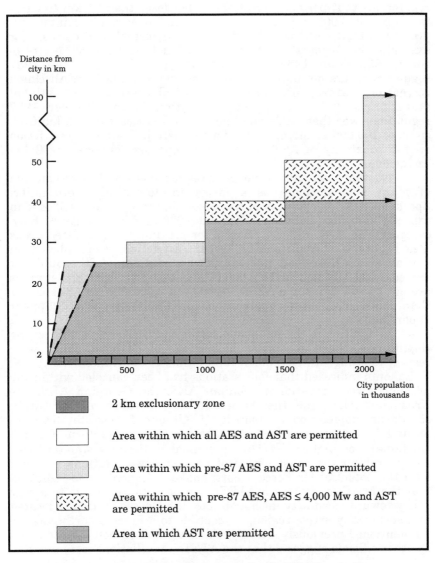

**Figure 4: Regulatory Parameters for Facility Proximity
to Population Centers, 1987-1990**

Sources: Sidroenko et al. "Normirovaniye bezopasnosti . . .," pp. 630; V. G.
Asmolov et al. "Rol' Reglamentiruyushchikh . . . , " pp. 461.

regulations that permitted stations to be closer than 25 km to cities with up to 300,000 in population.[34] As the Soviet media pointed out, previously built stations violated these newly established norms. For example, the Zaporozh'ye AES was 10 km from Nikopol', a city of some 160,000 in 1989.[35] Additionally, AESs under the 1987 regulations were not permitted to be less than 25 km from All-Union recreational areas, biosphere or historical preserves, or national state parks.[36] Another new restriction in the October 1987 regulations was that ASTs could not be located closer than 5 km from cities of 100,000 or more.[37] Previously there had been no restriction on the proximity of ASTs to a city of any size outside the 2 km exclusionary safety zone.[38]

Another departure from past regulations was the recognition of plant size or scale as a factor in determining proximity. Specifically, AESs were categorized by size, those up to 4,000 MW in capacity and those between 4,000 and 8,000 MW in capacity.[39] Each category was given different norms for proximity to populated areas (see Figure 4). Additionally, upper restrictions were placed on station size. Base-load nuclear powers stations (AES) could not exceed 8,000 MW and ASTs 2,000 MW in capacity.[40] As Soviet plans are not known to include AESs larger than 8,000 MW or ASTs larger than 2,000 MW, these restrictions on plant size would appear superfluous.

In addition to the published sections of the October 1987 siting regulations, there are other indications of changed policy with respect to siting. In December 1988, Minister of *Minatomenergo* N. F. Lukonin indicated that four stations had been canceled within the previous two years–the Azerbaijan AES, the Georgian AES, the Krasnodar AES, and the Armenian AES–because of new siting standards adopted about seismicity.[41] All four of these stations were located in regions acknowledged to be susceptible to a Force 7 earthquake or higher.*[42] Thus presumably, the new standards for siting prohibit the construction of nuclear power stations in areas with the potential for Force 7 earthquakes. Lukonin also indicated that the Minsk and Odessa ATETs had been canceled due to a new policy toward seismicity, although only the Odessa station is located in a seismically active region, susceptible to a Force 6 earthquake.[43] As mentioned previously, strong evidence based on policy statements by other officials suggests that the ATETs cancellations were due to a

---

* The Soviets use the MSK-64 scale of seismic intensity, which is also widely used in Europe. It is a 12-point scale like the modified Mercalli used in the United States. However, differences in the methods of measurement make these scales difficult to convert or compare.

reassessment of the ATETs configuation and a decision to improve reactor safety in this design.

Information Policy

As a result of the Chernobyl' accident, public confidence in nuclear industry decision-makers waned. This development and the expanded political freedoms under the Communist Party's new policy of Glasnost' combined to generate a considerable level of public opposition toward nuclear power by 1988. Nevertheless, Soviet industry officials and energy policy-makers were reluctant to alter established policies on public participation and access to information. In contrast to public policies concerning information and the Soviet public, the USSR made significant changes in its information policy with other countries. On both a bilateral and multilateral basis, the Soviet Union entered into several agreements allowing the exchange of information with other countries in the areas of reactor design, operation, and safety. Examples include Soviet reporting of reactor performance and daily operating incidents and accidents to the IAEA beginning in 1987.[44] In March 1988, regulatory institutions from the Soviet Union and the United Kingdom (the USSR State Committee for the Supervision of Nuclear Safety and the U.K.'s Nuclear Installations Inspectorate) agreed to establish a formal framework for the exchange of safety, related information on siting, construction, commissioning, and decommissioning of nuclear power facilities.[45] Soviet involvement in international information exchanges among nuclear industries progressed rapidly. By March 1989, the Soviet nuclear industry had joined WANO (World Association of Nuclear Operators) to develop an on-line computer system for the rapid exchange of information between reactor operators throughout the world.[46] Yet despite such horizontal information exchanges between national nuclear energy industries and national and international monitoring bodies, there was little headway in vertical information exchange between the Soviet nuclear industry and the Soviet public.

The reluctance of Soviet nuclear energy officials to release information on reactor safety and particularly site-related issues drew increasing fire from the Soviet media. For example, complaints were raised in January 1989 over official secrecy in the Soviet nuclear industry with regard to the Soviet public.[47] In particular, Soviet commentators explicitly contrasted domestic secrecy with the information policies toward the West in which Soviet officials shared detailed reactor and daily operational data with the

IAEA. Elsewhere in the press, complaints were leveled against ministerial secrecy, including secrecy over reactor location.[48]

Thus, as a result of the Chernobyl' accident, the Soviet nuclear power industry attempted to alleviate concerns about the nuclear industry's performance. However, despite concerted changes in reactor technology, reactor operation, safety and siting requirements, and to a very limited extent, institutional oversight, the Soviet nuclear power industry still maintained a monopoly on decision-making. The Soviet public still had no ability to challenge decisions made by the power ministry or to challenge the information used by the ministry to justify project decisions and industry policy. Nevertheless, as discussed in the next chapter, over time the Soviet public, particularly at the local level, did challenge decisions and was able to intervene in the decision-making process.

Notes

1. Theodore Shabad, "New Notes," *Soviet Geography*, Vol. 27, No. 7 (September, 1986), p. 504.

2. Ibid.; "Nuclear News Briefs," *Nuclear News*, August 1986, p. 29.

3. Ibid.

4. "A New Nuclear Ministry Has Been Formed in the USSR," *Nuclear News*, August 1989, p. 19.

5. Ibid.

6. Bukrinskii et al., p. 384.

7. V. G. Asmolov et al., "The Chernobyl' Accident: One Year Later," *Soviet Atomic Energy*, Vol. 64, No. 1 (July 1988), p. 21 (from *Atomnaya Energiya*, Vol. 64, No. 1, January 1988 p. 18–26).

8. Ibid.

9. "Chernobyl' Aftermath: Trees Recovering but Fallout of Faith Remains," *Nucleonics Week*, Vol. 30, No. 12 (March 23, 1989), pp. 6, 9.

10. Donahue et al., pp. 304–309; R. Gilette, "Soviet Reactor Badly Flawed Experts Say," *Los Angeles Times*, August 26, 1986, p. 6.

11. "No Feasible Alternative to Atomic Power," *CDSP*, Vol. 39, No. 17 (May 27, 1987), pp. 5, 9.

12. During the spring of 1988, Soviet reactor designers, among them Eugene Adamov, director of the Research and Development Institute of the GKAE, announced plans for the UKR-1500, an upgraded version of the RBMK-1500 with several passive safety features. Promotion of this reactor still based on the RBMK concept contradicted earlier announcements that the RBMK design was to be discontinued. "Soviets Report Development of 'Enhanced', Safer RBMK-1500," *Nucleonics Week*, Vol. 29, No. 12 (March 24, 1988), pp. 1, 10; "Soviets Developing Enhanced RBMK," *Nuclear Engineering International*, July 1988, p. 2.

13. "Construction of Chernobyl' Power Sets Cancelled," *FBIS*, FBIS SOV-89-076 (April 21, 1989), p. 90.

14. See Appendix A. Presumably this decision signaled the cancellation of Smolensk-4, Kursk-6, Chernobyl'-5 and -6, and Ignalina-3.

15. See Appendix A. 12,865 MW out of a total of 28,392 MW. This figure is based on the reactors in operation on January 1, 1986 and includes the VVER-1000, VVER-440, and the first-generation VVER-365 and VVER-210.

16. Troitskiy, *Energetika v SSSR*, p. 175.

17. Evidence suggests that since the late 1970s, reactor designers were constantly striving to improve the VVER design. Gold, pp. 46–50; G. A. Shasharin et al., "State-of-the-Art and Development Prospects for Nuclear Power Stations Containing Pressurized Water Reactors," *Soviet Atomic Energy*, Vol. 56, No. 6 (December 1984), pp. 365–366, (from *Atomnaya Energiya*, Vol. 56, No. 6, June 1984, pp. 277–281); F. Ya. Ovinnikov et al., "Opit Sozdaniya, ekspluatatsii i puti sovershenstvovaniya AES c VVER," in *Nuclear Power Experience*, Vol. 3 (Vienna: IAEA, 1982), pp. 51–67.

18. "Safety Record Reviewed on Large VVER Units," *Nuclear News*, January 1989, p. 103; Ye. I. Ignatenko, "Osnovniye napravleniya

rabot po povyshenko bezopasnosti seruynykh atomnykh energoblokov," *Elektricheskiye stanstii*, No. 8 (August 1989), p. 23.

19. "Safety Record Reviewed," p. 103; Ignatenko, p. 23.

20. Ignatenko, p. 23; V. A. Legasov and V. M. Novikov, "Bezopasnost' i effektivnost' yadernoi energetiki: kriteri, puti sovershenstvovaniya," in IAEA, *Nuclear Power Performance and Safety*, Vol. 2 (Vienna: IAEA, 1988), pp. 451–452.

21. "Chernobyl' Aftermath," p. 8.

22. "Seimens/GRS Get First Russian VVER Risk Assessment Contract," *Nucleonics Week*, Vol. 29, No. 39 (September 29, 1988), pp. 2–3; Bukrinskii et al., p. 384.

23. Petros'yants, *Atomnaya nauka*, pp. 64–66.

24. "USSR Seeks Public Acceptance of Improved Reactor Designs," *Nucleonics Week*, Vol. 29, No. 39 (September 29, 1988), pp. 1, 8–9.

25. Legasov and Novikov, pp. 451–452.

26. Kruglov, M. G. et al., "Prioritetiye napravleniya i gosudarstvenniye programmi nauchno-tekhnicheskogo progressa v proizvodstve i ispol'zovanii energeticheskykh resursov," *Teploenergetika*, Vol. 36, No. 1, (January 1987), p. 7.

27. "Germans and Soviets Agree to Cooperate on Modular HTRs," *Nucleonics Week*, Vol. 29, No. 43 (October 27, 1988), pp. 1, 8–9.

28. "No Feasible Alternative to Atomic Power," p. 5; "How Likely is Another Chernobyl'?" *CDSP*, Vol. 40, No. 42 (November 16, 1988), pp. 1–6; *Izvestiya*, November 27, 1989, p. 2.

29. Aleksashin et al., pp. 428–429.

30. "Industrial and Nuclear Power Safety Chairman Approved on the 14th," *BBC, SWB*, SU/0511 (July 18, 1989), C/8; V. G. Asmolov et al., "Rol' reglamentipuyushchykh polozheniy v povyshenii urovnya bezopasnosti atomnykh stantsiy," in *Regulatory Practices and Safety Standards for Nuclear Power Plants* (Vienna: IAEA, 1989), pp. 460–461.

31. "Atomic Power Minister Cited on Safety Measures," *FBIS*, FBIS SOV-88-027 (February 10, 1988), p. 71.

32. Asmolov et al., "Rol' reglamentipuyushchykh polozheniy," pp. 460–461.

33. Ibid.

34. Sidorenko et al., "Normirovaniye bezopastnosti," pp. 629–630.

35. "Industrial and Nuclear Power," *BBC, SWB,* C/8.

36. Asmolov et al., "Rol' reglamentipuyushchykh polozheniy," pp. 460–461.

37. Ibid.

38. Sidorenko et al., "Normirovaniye bezopastnosti," pp. 629–630.

39. Asmolov et al., "Rol' reglamentipuyushchykh polozheniy," pp. 460–461.

40. Ibid.

41. Other statements by Lukonin in the same month indicate that the decision to discontinue the Odessa ATETs was also a result of the new siting norms concerning seismicity. "In Armenia, Elsewhere," p. 67.

42. Charles K. Dodd, *Siting Hazardous Facilities in the Soviet Union: The Case of the Nuclear Power Industry,* Master's thesis, University of Washington, 1992, pp. 144–145.

43. "Minister Discusses Nuclear Power Development," *FBIS*, FBIS SOV-88-248 (December 27, 1988), p. 67.

44. IAEA, *Operating Experience with Nuclear Power Stations in Member States in 1990* (Vienna: IAEA, 1991).

45. "U.K./U.S.S.R.: Regulators Agree to Information Exchange," *Nucleonics Week*, Vol. 29, No. 14 (April 7, 1988), p. 11.

46. "Moscow Regional Center Reading as USSR Prepares for WANO Opening," *Nucleonics Week*, Vol. 30, No. 14 (March 9, 1989), pp. 3–4.

47. "Officials Challenged on Ecological Safety in Projects Near Kazan and Astrakhan," *BBC, SWB*, SU/0369 (January 27, 1989), C2/1.

48. *Pravda*, January 11, 1989, p. 3; *Izvestiya*, November 27, 1989, p. 2; *Izvestiya*, October 16, 1990, p. 6.

Chapter VI

New Participants in the Decision-Making Process: Public and Local Government Opposition

One result of the Chernobyl' accident was the emergence of widespread opposition among the general public to nuclear power in the Soviet Union. This opposition became increasingly vocal and organized over time, eventually influencing the nuclear power plans of central planners and power ministry officials. Additionally, there was a growing intolerance by local governments to accept the projects of energy planners and the nuclear industry. De facto, public and local opposition became a participant in the decision-making process, indeed to a limited extent even in the formal institutional process.

This chapter discusses the development of public and local government opposition to nuclear power in the wake of the Chernobyl' accident up to August 1991. The nature of this new public and local government opposition, its various institutional and social manifestations, and their net effects on nuclear power projects are discussed.

Public and Local Government Attitudes before Chernobyl'

Public or local government opposition to nuclear power in general or to individual projects was highly limited before 1986. In part, this was due to the extreme level of control the Communist Party and government exercised over Soviet society in the form of censorship, direct and indirect intimidation, as well as the limited avenues for political action on the part of opponents of established party policy. Another factor that explains the lack of discernable opposition or questioning of official policy was the lack of knowledge about the issues. This lack of knowledge on the part of the general public over issues in the nuclear industry undoubtedly contributed to the perceptions that there were few problems in this industry.

Nevertheless, public concern over nuclear power did indeed manifest itself before Chernobyl', particularly after the Three Mile Island accident in the United States in 1979. Public concern during this period was restricted to the form of letters of concern to local and national papers.[1] Most letters voiced apprehension about nuclear power plant safety. However, there were cases of local concern about the effects of a new nuclear station on the surrounding community. For example, in 1983 the Tatar paper *Kazan utlari* received many letters expressing misgivings over the potential environmental impacts (atmospheric releases of radiation and water pollution) from the planned Tatar AES as well as the potential demographic impacts, in particular the influx of skilled Russian labor to the new workers' city of Nizhnekamsk.[2] Concerns expressed via published letters were never presented as opposition, but rather were used as vehicles by the party to reassure the public of nuclear safety. Local party officials and the heads of the main construction organizations— *Kamgesenergostroy, Atomenergostroy,* and *Tatenergo*—assured the editors and readers of *Kazan utlari* via published notes from a special roundtable discussion that the effects of the new station would be minimal.[3]

The Soviet press frequently published articles and speeches by party officials and energy specialists supporting or advocating nuclear power in general, as well as individual projects. Occasionally, leading cultural figures supported local nuclear projects. For example, in 1981 the secretary of the Georgian Writers Union, Giorgi Tsitsishvili, and V. Gomelauri, Chairman of the Georgian SSR Academy of Sciences Council for Energy Problems, in a series of articles lobbied intensively for a base-load nuclear power station in the Georgian republic.[4]

Before 1986, local governments are not known to have opposed any nuclear power projects. Indeed local officials on several occasions appear to have appealed actively for nuclear power plants in their jurisdictions. For example, at the 26th Party Congress in March 1981, R. M. Musin of the Tatar ASSR pressed for accelerated construction of the Tatar AES.[5] On the eve of the Chernobyl' accident at the 27th Party Congress in February 1986, I. K. Polokov, of the Krasnodarskiy Kray brought attention to the fact that his region was experiencing severe electric power shortages and asked that the new five-year plan be altered to include the construction of a new nuclear power station in the Krasnodarskiy Kray.[6] Similarly, during February and March of 1986, the Party leadership of the Georgian SSR and the Azerbaijan SSR publicly supported nuclear stations in their republics.[7]

Local Opposition to Nuclear Power 1986–1991

Widespread public opposition to nuclear power surfaced soon after the Chernobyl' accident in 1986. This opposition, while a reaction to nuclear power in the wake of the catastrophic series of events at Chernobyl', was also a result of the loosening of the party's control of public discussion and public action in the form of the new series of policies collectively known as *glasnost'*. The Communist Party, under the leadership of Mikhail Gorbachev, used *glasnost'* to facilitate the economic reforms being undertaken, by creating avenues through which institutional waste, bureaucratic privilege, and official indifference to public welfare (i.e., *vedomstvennost'*— establishmentarianism or departmentalism) would be open to public discussion and, to a limited degree, action.   Intended to foster accountability to the public, *glasnost'* was reflected by greater investigative, critical, and combative reporting by the local, republican and national press; electoral reform increasing constituency power over local officials; and greater political automoumy by local authorities.

Under this new atmosphere of *glasnost'* the Soviet media and public became increasingly pugnacious. Indeed, between 1986 and 1991 public opposition to, and debate over, a number of industrial projects occurred in which nuclear power facilities represented only a subset.   Prominent among these other unpopular industrial projects were chemical production and processing plants, energy extraction projects, and hydroelectric stations.[8] In its initial stages, this "opposition" to nuclear power in the USSR did not take the form of a unified movement against the nuclear industry, rather it was manifested in a series of localized protests over the siting of a particular facility in a community or region.   In other words, opposition to nuclear power plants in the USSR took on a very "NIMBYish" character.

According to one Soviet source, between the Chernobyl' accident in April 1986 and July 1990, public opposition to plans for nuclear power stations had occurred at thirty-nine sites in the USSR.[9] Opposition to nuclear power projects appeared initially at only a few sites.   In 1987, opposition was publicly acknowledged concerning the Chigirin AES, the Odessa ATETs, and the Kharkov ATETs in the Ukrainian SSR; the Minsk ATETs in Belorussia; and the Krasnodar AES in the North Caucasus.[10] During 1988 opposition continued to spread, particularly in Ukraine, Belorussia and Lithuania (see Appendix 3). By 1989–1990 public opposition to nuclear projects intensified throughout the USSR as illustrated by the great increase in the number of sites involved; by the summer of 1990, more than twenty individual projects were subject to active and organized opposition.[11]

Issues

In each of the conflicts in which public protest occurred, the issue of the particular facility's location played a key role. This is not to say, however, that other issues were not the focus of concern. In many instances, reactor technology such as the continued use of the RBMK design (as at the Chernobyl' AES and the Ignalina AES), operator competence (as at the Kursk AES), and improper construction procedures (as at the Rostov AES) were key issues considered and expressed by local opposition.[12]

Among the site-related issues that attracted public attention, seismicity was a common concern at many sites. Seismicity refers to the risk of a damaging earthquake. At at least eight sites, there was a serious discrepancy between the estimates of power ministry experts and independent or outside experts about the likelihood of an earthquake or other forms of catastrophic geologic instability.[13] Many of the early protests focused on sites that were seismically questionable. The Krasnodar AES, Odessa ATETs, Crimean AES, Armenian AES, Georgian AES, Far East AES, Bashkir AES, and the Tatar AES were all stations were public opposition stemmed, at least in part, from seismic concerns (see Appendix C).

Another issue at many of the protested sites was the facility's proximity to densely populated areas. For technical and cost reasons, nuclear cogeneration stations (ATETs) and heat supply stations (AST) were to be located less than 20–40 km (12.4–24.8 miles) from the cities they were to supply.[14] For example, the Odessa ATETs site was 25 km (15.5 miles) from Odessa, the Gor'kiy AST approximately 12 km (7.4 miles) from the city center of Gor'kiy (Nizhne Novgorod), the Voronezh AST about 15 km (9.3 miles) from Voronezh, and the Arkhangel'sk AST about 5 km (3.1 miles) from Arkhangel'sk.[15] Between 1986 and 1990, some seven planned AST or ATETs sites were publicly opposed with station proximity playing a major role. At least two other planned stations, the Tatar AES and the Rostov AES, were opposed because of their close proximity to major population centers, Naberezhnyy Chelny and Nizhnekamsk in the case of the former, Volgodonsk in the case of the latter.[16]

Public concerns over potential ecological damage were the third and perhaps most common concern among opponents to nuclear power projects. At least fifteen station sites were opposed in part because of apprehension about potential radioactive releases and the consequential damage to the natural environment caused by plant operation. Examples of the former include the Chigirin AES, Tatar AES, Rostov AES, and Crimean AES on the Dnepr, Volga, and Don rivers and along the Black Sea coast. In each case, any contamination of local waters would have disastrous consequences for the environment downstream and for the large populations along these water bodies.[17] Other stations were opposed because of the thermal pollution in local water bodies and deleterious impacts of excessive and damaging levels of water withdrawals that would result from their planned capacities. Included among these stations were the Ignalina AES, the Chigirin AES, the South Ukraine AES, the Zaporozh'ye AES, the Tatar AES, and the Balakovo AES.[18]

Forms of Public Opposition

Public opposition to nuclear power in the Soviet Union took many forms between 1986 and 1991. From scattered incidents of public protest through letter writing, opposition became increasingly organized and militant. Moreover, as time went on, local opponents took advantage of many of the new quasi-democratic structures that were forming to influence energy and national policy-makers and to intervene in the decision-making process. Interestingly, throughout this time, public opposition largely retained its local character.

By far the most common form of protest, as well as the earliest to develop, took the form of public letters to the local and national media. Protest letters originated from three basic groups—public individual citizens and citizens groups writing public letters of protest; scientific experts and specialists publicly criticizing nuclear projects and the decision-making process via letters or articles in the media; and cultural figures, such as writers, using the media to inform the public of the potential ramifications of a project.

Public letters from individuals were widespread and presumably occurred in every case a project was opposed. Public letters and media articles presented by scientific experts played an extremely important role and were much more prominent in the regional and national press. Typically, these experts performed two functions. First, they informed the general public of the risks and potential impacts of a nuclear power station at a particular site. For example, Ukrainian geologists and other scientists brought attention to the seismic history of sites of the Crimean AES and the Odessa ATETs. Similarly, Soviet scientists objected to the Tatar AES and Bashkir AES due to their proximity to active faults.[19] Expert discussion on the ecological impact of thermal pollution and water withdrawals played a key role in the opposition to the Ignalina AES, Chigirin AES, Zaporozh'ye AES, Tatar AES, and Kostroma AES.[20]

Second, scientific experts, frequently proposed alternative plans or actions and critcized the established decision-making process. For example, economists argued for alternative modes of power production in the case of the Yaroslavl' ATETs and the Bashkir AES, arguing their case on economic grounds.[21] Another common proposal by experts was the establishment of the government or "expert" commission (*pravitel'stvennaya komissiya*).* These government commissions were to operate independently from *Minatomenergo* or later *Minatomenergoprom*. The purpose of the commission was to provide an objective, unbaised analysis of the site in question. Government commissions were requested at no fewer than twelve sites between April 1986 and December 1990 (see Appendix B). In most cases the government commissions, when requested, were formed. Typically, government commissions consisted of experts from many scientific fields and academic disciplines including geology, seismology, engineering economics, demography, and others.[22]

Cultural figures represented a third type of opposition by using the local and national media. Typically, this group included writers, literary figures, and historians. The role of cultural figures was particularly prominent in Ukraine where they led the opposition against the Chigirin AES, Crimean AES, and South Ukraine AES.[23] In each case, they brought the concerned nuclear projects to the forefront of the public's attention. For example, on August 6, 1987, a letter opposing the site of the proposed Chigirin AES was printed in *Literaturna Ukraina*. This letter was submitted by both cultural figures (writers and historians) and even local party officials. The letter criticized the proposed station on several grounds: the station would contribute to an already excessive level of water withdrawals from the Dnepr River; that local residents had not been consulted; and that the site is a well-known historical landmark in Ukraine, being the former capital of 17th-century Ukraine.[24] Similarly, a letter opposing the Crimean AES written by seven writers and scholars from Sevastopol' was published in the Kiev paper *Kul'tura i zhyitya*.[25] Ukrainian writers at the 19th Party Conference in June 1988 criticized the South Ukraine AES.[26]

While individuals, public groups, scientists, and cultural figures all used open letters through the media to express opposition, opposition manifested itself in other forms. Petition writing was frequently utilized by public opponents. Public petitions were reported at at least ten sites between 1986 and 1990.[27] These petition drives could be very substantial indeed. For example, in April 1989, more than 250,000 residents of the city of Nikolayaev signed a petition

---

* Literally "govenment commission", its usage usually connoted "expert" or "independent" commission.

opposing the South Ukraine AES; in July 1989, over 100,000 residents of Gor'kiy (Nizhniy Novgorod) signed a petition against the Gor'kiy AST; and in August 1990, more than 58,000 residents of Volgodonsk signed a petition opposing the construction of the Rostov AES.[28]

Demonstrations and protests were yet another form of public opposition adopted by residents and the local populace to oppose a project. At least ten projects became the object of this form of protest. This type of opposition was particularly common during the tumultuous months of 1989 and 1990. Protest rallies were held to show public opposition against the Ignalina AES in August 1988; the Tatar AES in April, July, and October 1989; the Crimean AES in October 1989; the Karelian AES in January 1990; the Balakovo AES and Khmel'nitskiy AES in June 1990; the Zaporozh'ye AES in July 1990; and the Rostov AES in August 1990.[29] Such protests varied of course in size, intensity, and media coverage. Early protests against the Ignalina AES attracted about 300 protesters in August 1988, whereas the protests against the Tatar AES in October 1989 involved several thousand people on a three-day march from Kazan University to the nuclear power station site at Kamskiye Polyany (a distance of about 100 km).[30] This latter protest was both televised and covered heavily in the national press. Protest rallies were typically organized by local public interest groups and occasionally by local politicians. For example, the August 1988 protests against the Ignalina AES were organized by the Lithuanian Restructuring Movement and the *Zemyna* ecology club, the June 1990 protests against the Balakovo AES by the Greens' Movement, the Zaporozh'ye AES by the local Greenpeace as well as People's Deputies from the surrounding oblast'.[31]

Even more extreme action was undertaken by individuals and organizations in the form of strikes or actual "blockades" of the nuclear station site. Such blockades were intended to disrupt ongoing construction work at the site and were usually a reaction to the passivity of higher authorities. In October 1989, local authorities and trade union organizations sanctioned protest strikes against the continuation of work on the Tatar AES and the Crimean AES.[32] In May 1990, in protest to the ongoing work to increase capacity at the Khmel'nitskiy AES, local enterprises refused to supply the plant with construction materials.[33] A month later, protesters blocked the flow of construction supplies to the Khmel'nitskiy AES.[34] Similar action was taken against the Rostov AES in August 1990 by local protesters.[35] In the case of the Tatar AES and the Crimean AES, strikes were called only after local authorities (i.e., ASSR and oblast' Soviets) had passed resolutions banning further construction (these resolutions were ignored by the USSR Council of Ministers and

Minatomenergoprom).[36] At the Khmel'nitskiy AES, protesters were responding to the local oblast' Soviet's failure to pass a moratorium banning further construction.[37] In the case of the Rostov AES, citizens instituted a daylong blockade after the USSR Council of Ministers refused to acknowledge an earlier resolution passed by rayon and oblast' authorities to halt work on the Rostov AES.[38]

Lastly, local public opposition used the local referendum, usually on the city or oblast' level, to protest decisions concerning nuclear power stations. Referendums were used at at least two sites to articulate opposition to the project. In May 1990, the city of Voronezh held a referendum on whether construction should continue at the Voronezh AST. With 81.5% of registered voters participating, more than 500,000 residents voted for cancellation (90% of those voting).[39] Earlier, in March 1990, a referendum was held in the city of Neftekamsk on the issue of whether the Bashkir AES should be canceled. Voter opinion against the project was almost unanimous with 99% of those who voted opposing the AES.[40] Significantly however, while the referendum underscored local public opposition to the Bashkir AES (Neftekamsk is 30 km from the station site), the Bashkir ASSR Supreme Soviet and All-Union bodies refused to recognize this referendum as official, because the process used was not stipulated in election legislation.[41]

Local Government Opposition

As mentioned previously in this chapter, evidence suggests that in many cases local government was an ardent supporter of power development projects including nuclear within their jurisdictions before 1986. Opposition from local government to the nuclear project plans of central authorities and *Minenergo* did not exist, at least publicly before 1986. After the Chernobyl' accident in 1986 however, this situation changed drastically. Initially, local government at the republican level (i.e., Union Republic and ASSR) responded to citizen demands for information and increased decision-making power by forming government commissions or by requesting All-Union bodies to establish government commissions to examine the site in question. Given the nature of the political reforms ongoing during 1986–1991 in the USSR, it was only a matter of time before public attitudes toward nuclear power would influence local governments' policies toward the same issues. By 1990 local governments throughout the USSR had attempted to intervene in projects decided upon and planned in Moscow. Significantly, local

Communist Party leaders began to openly question, criticize and denounce nuclear power projects in their localities. Local authorities in the Soviet Union were at first slow to respond to the public movements against nuclear power projects. However, in Lithuania, Belorussia, and Ukraine, the local Communist Party leadership and executive branches of government took the initiative in publicly questioning past decisions concerning nuclear power stations. The leadership of the Lithuanian Communist Party and the Lithuanian government openly criticized then-existing plans for the third RBMK-1500 reactor at the Ignalina AES.[42] In that same month—June 1988, Lithuanian authorities publicly requested the USSR Council of Ministers to resolve the problems and questions surrounding Ignalina's planned third reactor or to halt the project entirely.[43] In September 1988, the Belorussian Communist Party leadership and the Belorussian Council of Ministers requested the USSR Council of Ministers to cancel the highly controversial Minsk ATETs.[44] Similarly, the Vitebsk Oblast' Party First Secretary publically criticized plans for the planned Belorussian AES in a session of the Belorussian Supreme Soviet in November 1988.[45]

In Ukraine, in August 1987, the Communist Party First Secretary of the Poltava Oblast' criticized the proposed location of the planned Chigirin AES.[46] In February 1988, the Ukrainian Communist Party formed a special commission to reexamine nuclear power plant sites in Ukraine.[47] Further action by Ukrainian authorities followed in the fall of 1988 when the Ukrainian Communist Party and the Ukrainian Council of Ministers requested the USSR Council of Ministers to establish independent expert commissions to investigate the sites of the Crimean and South Ukraine AES (which were subsequently established in January 1989).[48] Even stronger action was taken in January 1989 when the Ukrainian Council of Ministers requested All-Union authorities to halt construction on the Chigirin AES.[49]

Local government in the Soviet Union took on a new complexion after nationwide elections for local executive and legislative positions during the spring of 1989. Many candidates ran on environmental platforms, including those that opposed nuclear power. As a result, 1989 was marked by a considerable increase in local government opposition to nuclear power facilities. In September, the Crimean Oblast' Soviet passed a resolution banning further work on the Crimean AES and requesting the USSR Council of Ministers to formally cancel the project.[50] This set the pattern that would be become quite common over the next two years. Legislative bodies at the city and oblast' level would pass resolutions banning any further work or postponing work until independent investigation of the site had been completed. Formally, this had little effect because the USSR Council of Ministers was the only government body that could cancel a ministry's project. Thus, most local government resolutions banning further work also included appeals to the USSR Council of Ministers for a final decision on the project. At nine other sites in question between September 1989 and December 1990, local oblast' or city Soviets passed resolutions banning further work at the site and requesting the USSR Council of Ministers to cancel the project (see Table 16).

Table 16: Local Governments Requesting the Cancellation of Nuclear Power Stations, 1986–1991

Station	Date	Political Unit	Institution
Minsk ATETs	9/88	Belorussian SSR	Belorussian SSR CP, Belorussan SSR C-of-M
Chigirin AES	1/89	Ukrainian SSR	Ukrainian SSR C-of-M
Yaroslavl' ATETs	5/89	Yaroslavl' Oblast'	Yaroslavl' Oblast' CP
Crimean AES	9/89	Crimean Oblast'	Crimean Oblast' CP, Crimean Oblast' SS
Arkhangel'sk AST	2/90	Arkhangel'sk Oblast'	Arkhangel'sk Oblast' Executive Committee, Arkhangel'sk Oblast' SPD
Karelian AES	3/90	Karelian ASSR	Karelian ASSR SS
Perm AES	4/90	Perm Oblast'	Perm Oblast' Soviet
Gor'kiy AST	5/90	City of Gor'kiy	Gor'kiy City Soviet
Tatar AES	6/90	Tatar ASSR	Tatar ASSR SS
Gor'kiy AST	8/90	Gor'kiy Oblast'	Gor'kiy Oblast' SPD
Bashkir AES	9/90	Bashkir ASSR	Bashkir ASSR SPD

Note: CP denotes Central Committee Communist Party, C-of-M denotes Council of Ministers, SS denotes Supreme Soviet, SPD denotes Soviet of People's Deputies.

Sources: See Appendix C.

Local governments worked in other ways to oppose nuclear power projects. For example, in June 1989, the Communist Party leadership in the Tatar ASSR and the Tatar ASSR Council of Ministers requested the USSR Council of Ministers to form an government commission to investigate the Tatar AES.[51] In Ukraine, the controversy over the siting of the Crimean AES led to perhaps the most audacious series of acts by local government up to that time. In October 1988, a Ukrainian government commission (formed earlier that same year at the request of the Ukrainian Communist Party and Ukrainian Council of Ministers) ruled against the proposed Crimean AES site because of the seismic and volcanic risk at the site.[52] Nevertheless, both the *Minatomenergoprom*, as well as the ultimate economic authority, the USSR Council of Ministers, failed to act upon the request for a cessation of construction.[53] Rather, the USSR Council of Ministers voted to continue investigation into the seismic characteristics of the site while construction continued.[54] After twelve months local authorities, reacting to widespread public sentiment in the Crimea against the station, decided to take matters into their own hands. In October 1989, the Crimean Oblast' Soviet of People's Deputies passed a resolution demanding that the USSR Council of Ministers halt construction on the Crimean AES. The Crimean Oblast' Soviet took the additional measure of instructing the oblast' Industrial Construction Bank (*Prostroybank*) to cut off funding for the work.[55] Additionally, the chairman of the Crimean Oblast' Trade Union Council stated that if the USSR government continued to avoid making a decision on the AES, then the oblast' Soviet would sanction warning strikes by Crimean enterprises protesting continued construction.[56] Central authorities responded quickly, with the USSR Council of Ministers announcing the project's cancellation on October 27, 1989.[57]

By 1990 the Soviet Union had entered a new phase in relations between central and local governments. Local governments, particularly the republics, attempted to exercise greater sovereignty in a number of issue areas including energy policy and decisions concerning nuclear power. On March 1, 1990, the Ukrainian Supreme Soviet enacted comprehensive legislation on the protection of the Ukrainian environment.[58] Among the measures enacted was a provision ordering the shutdown of the Chernobyl' AES by 1995 and a cessation of any additional construction at the Rovno AES and the Khmel'nitskiy AES.[59] The Ukrainian republic took even stronger measures in August of that year. After considerable public and local government opposition during the spring and summer of 1990, the Ukrainian Supreme Soviet on August 2 declared a five-year moratorium on the commissioning of new nuclear power stations and the expansion of capacity at existing plants in the Ukrainian republic.[60] This legislation had followed the republic's declaration of state sovereignty on July 16, 1990.[61] The republic of Estonia followed with similar legislation in November when the republic's national legislature suspended planning and construction work on the planned Estonian AES until the year 2000.[62]

Local Opposition and Its Effect on the Soviet Nuclear Industry

The eventual scale and intensity of local public and local government opposition to existing nuclear power projects led to unprecedented consequences for the Soviet nuclear industry and ultimately the energy sector. What resulted was a massive hamstringing of an industry slated for most of the planned growth for what was traditionally one of the highest priority areas in the Soviet national economy—electric power generation.[63] Indeed, between 1986 and 1990, some thirty-nine nuclear projects were canceled or indefinitely postponed in the USSR due directly or indirectly to public opposition.[64] Of these thirty-nine projects, four were capacity expansions at existing sites, fifteen were projects in the construction stage, and twenty were at the planning or site analysis stage.[65] According to another Soviet source, this amounted to approximately 100,000 MW of planned power generation capacity.[66] The monetary losses from these cancellations were, needless to say, very substantial. In terms of ruble investments foregone (i.e., investment outlays in projects that were subsequently canceled), these losses amounted to a total exceeding 5 billion rubles.[67]

As a result of the Chernobyl' accident and relatively open political forum created under *glasnost'*, the Soviet public came to question policies, as well as the legitimacy of decision-making in the Soviet nuclear power industry. This opposition, generally locally inspired and locally based, employed a number of strategies that worked through established institutions, many of which were new in the wake of political reforms sweeping the Soviet Union between 1988 and 1991. Additionally, however, frustrated opponents often resorted to confrontational strategies working outside established institutions and political channels. The effectiveness of local opposition is perhaps surprising given the lack of formal institutional processes for public oversight or participation in decision-making in the nuclear industry. Nevertheless, it is clear that local communities and governments were able to influence the ultimate fate of nuclear power projects.

Notes

1. For example, see *Komunsti* (in Georgian), February 18, 1981; *Trud*, June 5, 1981; *Trud*, September 29, 1982; Sergei Voronitsyn, "Further Debate on the Safety of Nuclear Power Stations in the USSR," *RFE / RL, Radio Liberty Research Reports*, RL 350/81 (September 7, 1981), pp. 3–4; Sergei Voronitsyn, "How Great is Soviet Citizen's Fear of Nuclear Radiation," *RFE / RL Radio Liberty Research Reports*, RL 468/82 (November 22, 1982), pp. 1–2; Voronitsyn, "Concern in Tatar ASSR," pp. 2–3.

2. Voronitsyn, "How Great is Soviet," pp. 2–3.

3. Ibid.

4. Elizabeth Fuller, "Is Nuclear Power the Answer to Georgia's Energy Problems?" *RFE / RL, Radio Liberty Research Reports*, RL 282/81 (July 17, 1981), pp. 1–3.

5. Donna Bahry, *Outside Moscow: Power, Politics, and Budgetary Policy in the Soviet Republics* (New York: Columbia University 1987), p. 140.

6. *Pravda*, March 1, 1986, p. 2.

7. "Krasnodar Public Opinion That Caused Cancelling of Nuclear Power Station Laid to Uninformed Post–Chernobyl' Fear," *CDSP*, Vol. 40, No. 3 (Febuary 19, 1988), p. 9 (from *Komsomol'-skaya Pravda*, January 27, 1988 p.4).

8. Ann Sheehy and Sergei Voronitsyn, "Ecological Protest in the USSR, 1986–1988," *RFE / RL, Radio Liberty Research Reports*, RL 191/88 (May 11, 1988), pp. 1–3.

9. "Chelyabinsk Nuclear Power Station Planned," *FBIS*, FBIS SOV-90-238 (December 11, 1990), p. 69. The author was able to indentify thirty-three of these.

10. See Appendix C.

11. By 1989 protests had spread throughout the RSFSR including to the Far North, the Volga Basin, the Don Basin, the Urals, and the Far East (see Appendix C).

12. Girnius, "Continued Controversy," p. 30; *Izvestiya*, August 2, 1990, p. 2; *Izvestiya*, August 11, 1990, p. 2.

13. See Appendix C.

14. Tokarev, p. 201.

15. V. A. Tsygankov et. al., "Atomnaya stantsiya teplosnabzheniya dlya otdalennykh rayonov," *Teploenergetika*, No. 12 (December 1981), p. 9; *Komsomol'skaya Pravda*, May 24, 1990, p. 2; "Work Halted on Arkhangel'sk Nuclear Power Plant," *FBIS*, FBIS-SOV-90-035 (February 21, 1990), pp. 114–115.

16. Reportedly, the Tatar AES was 30 km from Naberezhnyy Chelny and 32 km from Nizhnekamsk, and the Rostov AES 13 km from Volgodonsk. "Tatar Anti-Nuclear Protests Reported," *FBIS*, FBIS-SOV-89-146 (August 1, 1989), p. 79; "Workers Protest Jobs at Rostov Nuclear Plant," FBIS, FBIS-SOV-90-220 (November 14, 1990), p. 45.

17. David Marples, "Chigirin and the Soviet Nuclear Energy Program," *RFE/RL*, RL 366/89 (July 9, 1989), pp. 26–29; Roman Solchanyk, "Ukrainians Send Appeal on Nuclear Energy to Party Conference," *RFE/RL, Radio Liberty Research Reports*, RL 294/88 (June 29, 1988), p. 4; *Pravda*, January 11, 1989, p. 3; "Construction of Bashkiriya AES Halted," *FBIS*, FBIS-SOV-90-171, (September 4, 1990), p. 106.

18. Marples, "Chigirin and the Soviet Nuclear," pp. 26–29; Girius, "Continued Controversy," pp. 32; Solchanyk, "Ukrainians Send Appeal," p. 4; *Pravda*, October 5, 1989, p. 6; *Izvestiya*, August 11, 1990, p. 2; "Ukrainian Nuclear Plant Picketed," *FBIS*, FBIS-SOV-90-113, (June 12, 1990), p. 114; "Protest against Karelian Power Plant Construction," *FBIS*, FBIS-SOV-90-012 (January 18, 1990), p. 132; "Balakovo Protests against Nuclear Power Station," *FBIS*, FBIS-SOV-90-113 (June 12, 1990), p. 109; *Sovetskaya Rossiya*, April 26, 1989, p. 3; "Tatar Anti-Nuclear Protests Reported," p. 80; "Deputy Minister on Public Concern over AES Contamination of Waterways," *JPRS*, JPRS-UPA-89-045 (July 20, 1989), pp. 44–46.

19. "Tatar Anti-Nuclear Protest Reported," p. 80; *Pravda*, January 11, 1989, p. 3; "Officials Challenged," *BBC, SWB*, SU/0369

(January 27, 1989), C2/1; "Construction of Bashkiriya AES Halted," *FBIS*, FBIS-SOV-90-171 (September 4, 1990), p. 106.

20. Bohdan Nahaylo, "More Ukrainian Scientists Voice Opposition to Expansion of Nuclear Energy Program," *RFE/RL, Radio Liberty Research Reports*, RL 135/88 (March 21, 1988), pp. 1–2; Saulius Girnius, "The Ignalina Atomic Plant's Second Reactor in Operation," *RFE/RL, Radio Liberty: Baltic Area*, SR/4 (May 4, 1988), pp. 12–13; "Officials Challenged on Ecological Safety in Projects near Kazan and Astrakhan," *BBC, SWB*, SU/0369 (January 27, 1989), C2/1; "Industrial and Nuclear," *BBC, SWB*, C/8; "Kostroma Atomic Power Plant Construction Halted," *FBIS*, FBIS-SOV-90-130 (July 6, 1990), p. 62.

21. *Sovetskaya rossiya*, May 21, 1990, p. 3; "Construction of Bashkiriya," *FBIS*, September 4, 1990, p. 106.

22. *Izvestiya*, October 27, 1989, p. 3.

23. Marples, "Chigirin," pp. 26–29; Solchanyk, "Ukrainian Writers Protest," pp. 1–2; Solchanyk, "Ukrainians Send Appeal," p. 4; *Literaturna ukraina*, June 23, 1988, p. 4.

24. Marples, "Chigirin," pp. 26–29; Solchanyk, "Ukrainian Writers Protest," pp. 1–2. The letter in its critcism of *Minenergo*'s selection of the site for the Chigirin AES, descibes a rather remarkable site selection process. According to the dissenting Ukrainian writers, in 1969 *Minenergo* announced that it planned to build a 4,800 MW coal-fueled GRES on the shore of the Dnepr Reservoir at the site of the later Chigirin AES. Eventually, after several million rubles had been invested in the project, the plans were altered in favor of a 1,600 MW oil-fueled GRES. During the mid-1980s after further construction at the site, it was realized that a fossil-fueled GRES was not viable because the requisite fossil fuel for the station was not economically available. *Minenergo* decided that a 2,000 MW nuclear plant was a economic substitute. *Minenergo* officials appearently made this decision in 1985 without obtaining official clearance through the normal siting process.

25. Solchanyk, "Ukrainians Send Appeal," p. 4.

26. Nahaylo, pp. 1–2.

27. See Appendix B.

28. *Izvestiya*, October 28, 1989, p. 3; "100,000 Sign Nuclear Power Plant Protest Message," *FBIS*, FBIS-SOV-89-146 (August 1, 1989), p. 81; "From the Stenograph of the Congress of People's Deputies," *FBIS*, FBIS-SOV-89-142-S (July 26, 1989), p. 21.

29. *Pravda*, October 5, 1989, p. 6; *Izvestiya*, August 11, 1990, p. 2; Girnius, "Continued Controversy," p. 32; "Ukrainian Nuclear Plant Picketed," p. 114; *Sovetskaya rossiya*, April 26, 1989, p. 3; "Balakovo Protests against Nuclear Power Station," *FBIS*, FBIS SOV-90-113 (June 12, 1990), p. 109; "Protest against Karelian Power Plant Construction," *FBIS*, FBIS-SOV-90-012 (January 18, 1990), p. 132.

30. Girnius, "Continued Controversy," p. 32; *Pravda*, October 5, 1989, p. 6.

31. Girnius, "Continued Controversy," p. 32; "Balakovo Protests," p. 109; "Ukrainian Nuclear Plant," p. 114.

32. *Sovetskaya rossiya*, October 22, 1989, p. 1; *Sotsialisticheskaya industriya*, October 21, 1989, p. 1.

33. "Khmelnitskiy Holds Ecological Meeting–May 27," *FBIS*, FBIS SOV-90-130 (July 6, 1990), p. 68.

34. "Pickets Block Khmelnitskaya Atomic Power Station," *FBIS*, FBIS-SOV-90-140 (July 20, 1990), p. 100.

35. *Izvestiya*, August 11, 1990, p. 2; "Bashkiria Supreme Soviet Debates Nuclear Plant," *FBIS*, FBIS-SOV-90-172 (September 5, 1990), p. 81.

36. *Sovetskaya rossiya*, October 22, 1989, p. 1; *Sotsialisticheskaya industriya*, October 21, 1989, p. 1.

37. "Pickets Block Khmelnitskaya," p. 100.

38. *Izvestiya*, August 11, 1990, p. 2.

39. *Komsomolskaya pravda*, May 24, 1990, p. 2.

40. "Neftekamsk 'Referendum' Criticizes Bashkir AES," *FBIS*, FBIS-SOV-90-079-S (April 24, 1990), p. 39.

41. Ibid.

42. Girnius, "Continued Controversy," p. 30.

43. Ibid.

44. *Izvestiya*, September 7, 1988, p. 2.

45. Sagers, "News Notes," (April 1989), p. 340.

46. Solchanyk, "Ukrainian Writers Protest," p. 1.

47. Roman Solchanyk, "More Controversy on Nuclear Energy in the Ukraine," *RFE / RL, Radio Liberty Research Reports,* RL 231/88 (June 8, 1988), p. 3.

48. "Ukraine Environment, Nuclear Plants Debated," *FBIS*, FBIS-SOV-89-014 (January 24, 1989), p. 71.

49. Ibid.

50. *Sotsialisticheskaya industriya*, October 21, 1989, p. 1.

51. "Marchers Oppose Tatar AES," *FBIS*, FBIS-SOV-89-208 (October 30, 1989), p. 80.

52. *Pravda*, January 11, 1989, p. 3.

53. *Sotsialisticheskaya industriya*, October 21, 1989, p. 1.

54. Ibid.

55. Ibid.

56. Ibid.

57. *Izvestiya*, October 27, 1989, p. 3.

58. David Marples, "Decree on Ecology Adopted in Ukraine," *RFE / RL, Report on the USSR*, RL 161/90 (March 15, 1990), p. 15.

59. Ibid., p. 16.

60. David Marples, "Ukraine Declares Moratorium on New Nuclear Reactors," *RFE/RL, Report on the USSR,* RL 425/90 (August 20, 1990), p. 20.

61. Ibid.

62. Marples, "Ukraine Declares Moratorium," p. 20.

63. As mentioned in Chapter II, during the 11th Five-Year Plan (1981–1985), planned nuclear capacity additions amounted to 36.3% of total planned capacity additions. For the 12th Five-Year Plan (1986–1990), planned nuclear capacity additions amounted to 47.5% of total planned capacity additions.

64. *Pravda,* July 19, 1990, p. 2.

65. Ibid.

66. "Chelyabinsk Nuclear Power Station," p. 69.

67. Ibid.

Chapter VII

The Legacy of Soviet Decision-Making and Nuclear Power

This study examined the issue of project planning and facility siting in the former Soviet Union, focusing on participating institutions in the decision-making process, policies relating to location, scale, and technology, and the resulting conflicts between central authorities and local interests. During the period previously discussed, decision-making in the Soviet nuclear power industry was highly centralized and concentrated in the traditional organs of Soviet economic administration. Oversight or regulatory institutions serving the interests of the local public were nonexistent. Those oversight and regulatory institutions that did exist were largely tied to the interests of the power ministry. As a result, over time the Soviet nuclear power industry developed at a rapid pace, with cost reduction and speed of construction primary considerations for decision-makers. As a consequence, policies encompassing location, scale, and technology tended to overlook safety issues. After the Chernobyl' accident, heightened public concern over the safety of nuclear power in the USSR, combined with ongoing political reforms led to direct conflicts between the local public and local government and the various institutions involved in the administration of the electric power industry. As local interests appeared to be gaining noticeable influence over previously centrally directed projects, an event of monumentous proportions occurred—the political disintegration of the Soviet Union into a collection of independent states.

Developments in the Newly Independent States of the Former USSR

The failure of the August 1991 Coup by hardliners in the Communist Party leadership marked an important watershed in the relationship between local government and Moscow, particularly at the republican level, and essentially crowned centrifugal trends that had been occurring over the preceding three years. The net effect from the political fallout of the August Coup has been a divergence of institutions and policy within the newly independent states of the former Soviet Union.

Most of the newly independent states of the former Soviet Union, on achieving independence, have entered into a loose confederation known as the Commonwealth of Independent States (CIS). Other states including the Baltic states—Lithuania, Latvia and Estonia, as well as Georgia—have maintained near complete political independence, resolving trade, environment, and military issues bilaterally with their former Soviet neighbors. The CIS essentially constitutes a quasi-confederation intended as a framework for resolving issues arising from the degree of military and economic integration between the former republics of the Soviet Union. Nowhere is the issue of regional or former republican economic interdependence more pronounced than in the energy sector. Former Soviet republics were highly interdependent on each other for conventional primary fuels such as coal, oil, and gas; the production and distribution of electric power; and within nuclear power, reactor design, nuclear fuel production, and nuclear waste processing and disposal. Not only has the breakup of the former Soviet Union interrupted the unified pricing, transport, and delivery of these goods as well as the supply of key inputs for their production, extraction, or distribution it has also changed former institutions managing and regulating those activities.

The breakup of the USSR contributed to the further deterioration of the energy sector that had begun in the late 1980s. The production of oil, gas, and coal in the newly independent states continued to fall, due inpart to long-term tends discussed in Chapter II and particularly after the August Coup, declining capital investment, interrupted supply of key inputs, and increasing labor unrest. Indeed, by the end of December 1991 (the USSR finally ceased to exsist on December 25, 1991), all major industries of the Soviet energy sector including enterprises in states that had claimed nominal independence, such as Lithuania, reported declines in total output. Coal production fell by 10.5%, oil production dropped by 9.8%, gas by 0.5%, and electricity by 2.9% or 50 billion kWh (to a total of 1676 billion kWh for 1991), the first decline for electric power since the Second World War.[1]

The new nations of the Russian Federation, Ukraine, Lithuania, Kazakhstan, and Armenia all inherited nuclear plants from the former Soviet nuclear power industry, as well as the responsibility for their operation and safety. The Russian Federation by far maintains the largest nuclear industry in the CIS, with all the major elements of plant design, operation, and even the fuel cycle under its authority. As of July 1992, Russia operated twenty-four commercial reactors at nine sites (see Appendix D).[2] Nuclear power accounts for approximately 11% of total electrical generation in the Russian Federation.[3] However, as one might expect nuclear power plays an even greater role in European Russia; St. Petersburg relies on nuclear power for 33% of its electricity supply and the Moscow region for 22%.[4]

Since the formation of the CIS, the nuclear industry in the Russian Federation has been significantly reorganized. On April 1, 1992, the Russian Ministry of Atomic Energy (*Minatom*) came into existance replacing *Minatomenergoprom*. *Minatom* authority apparently extends over research, plant design, construction, and operation, as well as the fuel cycle. Structurally within *Minatom*, a consortium of nuclear power plants known as *Rosenergoatom* operates the country's nuclear plants.[*5] The Nuclear and Radiation Safety Inspectorate (*Gosatomnadzor* RF)[**] has replaced *Gospromatomnadzor* as the industry's main safety regulatory body.[6] *Gosatomnadzor* RF and the State Health and Epidemiological Inspectorate (*Sanepidemnadzor* RF) are now responsible for licensing and regulation of nuclear power plants, fuel cycle facilities, and factories in the Russian nuclear power industry.[7]

Nuclear policy in the Russian Federation is still evolving in the realms of siting, scale, and technology. Announcements by industry officials indicate tacit recognition of the authority of local government bodies and public acceptance in any future siting decisions.[8] The role of the ministry in the siting process is to conduct feasibility studies and suggest options to local governments for approval. Indeed, the scale of planned expansion has been reduced, although the First Deputy Prime Minister of the Russian government instructed the government to increase capacity at eleven existing stations and new sites in March 1992.[9] A serious accident at the Leningrad AES on March 24, 1992, only increased anxiety about the Russian nuclear industry, particularly the use of the RBMK design reactors.[***] By June, nuclear officials had admitted they plan to add 9,000 MW of

---

[*] The Leningrad AES (recently renamed the Sosnovy Bor AES) is not a member of *Rosenergoatom* and operates independently.

[**] To avoid confusion with the former Soviet institutions, of the same name RF is used to denote the new institutions of the Russian Federation.

[***] This accident involved the Leningrad 3 reactor on March 24, 1992. A control valve broke, blocking the flow of coolant to a fuel channel ultimately causing damage to the fuel. The accident brought increased attention to inadequate safety design of the first-generation RBMKs ("Leningrad-3 Channel Damage Due to Control Valve Failure," *Nucleonics Week*, Vol. 33, No. 14 ,April 2, 1992, pp. 1–2).

capacity by the year 2000. As of July 1992, *Minatom* was conducting feasibility studies at six new sites.[10]

In December 1992, the Russian government announced a revised program for the Russian Federation's nuclear power industry. This program represents an ambitious revival of the Russian nuclear industry with several, albeit modified, elements of old Soviet plans. The program calls for the addition of approximately 16,500 MW of nuclear capacity before 2010–2015.[11] The program includes a plethora of modified designs including one modified channel-type, graphite-moderated reactor based on the RBMK design; three graphite-moderated, light-water reactors; four fast-breeder reactors; and fifteen pressurized-water reactors based on the VVER design (see Table 17).[12] The new plan is notable in several respects: the continuation of the RBMK with the startup of the nearly complete Kursk-5 unit, the continuation of controversial ATETs and AST projects for district heating, the diversification in reactor designs, and relatively few new station sites although three of these (the planned Far East AES, Primorskaya AES, and Khabarovsk AST) are located in the Russian Far East. The plan also includes the decommissioning of Novovoronezh-3 and -4 and Kola-1 and -2, the only remaining first-generation VVER-440s. *Minatom* officials insist that these projects must be accepted by local authorites before construction begins. Local authorities have formally requested the completion of reactor additions at Balakovo, Kalinin, and Kursk; and local governments in the Far East and in St. Petersburg (with jurisdiction over Sosnovy Bor) have recently given support for projects in their areas.[13]

The Republic of Ukraine also was bequethed a considerable portion of the former Soviet nuclear power industry. However, the Ukrainian industry is dominated by power production and lacks the reactor design and particularly the fuel extraction, processing, and waste disposal elements of the fuel cycle. Fourteen commercial reactors at five sites operate in Ukraine, supplying 27% of Ukrainian electricity in 1992.[14] Nuclear plants in Ukraine have been organized into an operating consortium known as *Ukratomenergoprom*. The primary safety organization in Ukraine is the State Committee for Nuclear and Radiation Safety (*Derzhatomnaglyad*), which replaced the Moscow-based *Gospromatomnadzor*.[15]

Nuclear policy is still being formulated in Ukraine and by the summer of 1992 had become a battleground between the executive and legislative branches of that nation's government. As of

Table 17: Revised Nuclear Construction Program for the Russian Federation, December 1992

Station	Reactor Type	Capacity (MW)	Unit Name	Start-up Date
Balakovo AES	VVER-1000	1,000	Balakovo-4	1993
Kalinin AES	VVER-1000	1,000	Kalinin-3	1995
Kursk AES	RBMK-1000 (imp.)	1,000	Kursk-5	1995
Bilibino ATETs	GLWR	32	Bilbino-5	2001–2005
	GLWR	32	Bilibino-6	2001–2005
	GLWR	32	Bilibino-7	2001–2005
South Urals AES	BN-800	800	South Urals-1	2000
	BN-800	800	South Urals-2	2000
	BN-800	800	South Urals-3	after 2000
Beloyarsk AES	BN-800	800	Beloyarsk-4	after 2000
Novovoronezh AES	VVER-1000 (imp.)	1,000	Novovoronezh-6	2001–2005
	VVER-1000 (imp)	1,000	Novovoronezh-7	2001–2005
Kola AES	VVER-630	630	Kola-5	2001–2005
	VVER-630	630	Kola-6	2001–2005
	VVER-630	630	Kola-7	2006–2010
Far East AES	VVER-600	600	Far East-1	2010
	VVER-600	600	Far East-2	2010
Primorskaya AES	VVER-600	600	Primorskaya-1	2010
	VVER-600	600	Primorskaya-2	2010
Voronezh AST	AST-500	500	Voronezh-1	2000
	AST-500	500	Voronezh-2	2000
Khabarovsk AST	AST-500	500	Khabarovsk-1	2010
	AST-500	500	Kharbarovsk-2	2010
Sosnovy Bor AES	Pilot V-630	630	Sosnovy Bor-5	2010
	Pilot V-630	630	Sosnovy Bor-6	2010

Note: (imp.) denotes enhanced safety version of the reactor type. Sosnovy Bor AES refers to the former Leningrad AES, which underwent a name change in 1992.

Source: "Russia Okays Plan to Proceed with Major Nuclear Construction," *Nucleonics Week*, Vol. 34, No. 3 (January 21, 1993) pp. 12–13.

September 1992, the Ukrainian government had no new projects for the nuclear industry. However, officials at *Ukratomenergoprom* plan to bring into operation the almost completed reactors at the Zaporozh'ye AES, Rovno AES, and Khmel'nitskiy AES to replace capacity lost when the remaining reactors are shut down at Chernobyl' by the end of 1993. Bringing these reactors into operation is still highly problematical given the local opposition to capacity expansion at these sites. The issue of the still operating Chernobyl' reactors has been fiercely contested. Before the August Coup, the Ukrainian parliment had voted to shut down all the Chernobyl' reactors by the end of 1993. On May 25, 1992 in the wake of the accident at the Leningrad AES, the Ukrianian Parliment voted, and *Derzhatomnaglyad* recommended, to shut down permanently all reactors at the Chernobyl' station. Nevertheless, this decision was overruled by Ukrainian President Leonid Kravchuk who ordered a restart of these reactors in October 1992 upon completion of safety upgrades on reactor control valves.[16] In October the Ukrainian legislature, concerned over possible inadequacies of a safety regulatory agency (*Derzhatomnaglyad*) funded by the executive branch of the Ukrainian government, has moved to extend the powers of the Parliament's Commission for the Chernobyl' Catastrophe to the oversight of the entire nuclear industry.[17]

Upon becoming independent, the government of Lithuania took over the two 1,500 MW reactors at Ignalina. Of all the newly independent states of the former Soviet Union, Lithuania is the most dependent upon nuclear power, deriving approximately 55% of its electricity from these two reactors.[18] Ignalina is under the operation of the Lithuanian Ministry of Energy. Regulatory oversight is administered by the Nuclear Safety Inspectorate (VATESI), which is independent of the Ministry of Energy.[19] Lithuanian government and industry officials do not intend to increase existing capacity. However, the Ministry of Energy is interested in eventually replacing Ignalina-1 and -2 with a new, safer design of Russian or Western origin at the same site.

Kazakhstan inherited the 350 MW commercial breeder reactor at Shevchenko, as well as several industrial enterprises crucial to the fuel cycle in the old Soviet nuclear power industry. In Febuary 1992, Kazakhstan created the Kazakh State Atomic Power Engineering and Industry Corporation (KATEP). KATEP operations include production of uranium and other minerals, nuclear project management, and construction, in addition to forming commercial partnerships for those projects with foreign firms.[20] The formation of KATEP was followed by the creation of a safety regulatory institution—the Kazakh Atomic Energy Agency. The Kazakh Atomic Energy Agency is a cabinet-level agency responsible for nuclear safety regulation and nuclear export controls and transport. The Kazakh nuclear power industry is to continue its role as a supplier of plutonium and maintain operation of the Shevchenko reactor until 2003, when KATEP intends to replace the existing reactor with an improved 350 MW fast-breeder design at the same site.[21]

Since the formal independence of Armenia, there has been some discussion of restarting the two 440 MW reactors in Armenia which were closed down in 1989. In January 1992, the Armenian government announced that was readying the Armenia-2 reactor for restart by the end of 1992.[22] This was a surprising reversal of the sentiment of the local authories who during 1988–1989 opposed the station because of the poor safety design of the first-generation VVER-440, its controversial location in a seismic region, and continued public opposition. However, the political conflict between Armenia and Azerbaijan has made the Russian Federation reluctant to provide the technical assistance and fuel necessary for startup. Despite the announcements by the Armenian political and government leadership of their intention to startup the Armenian-2 reactor, little has occurred to make this a reality.

Changing Local Attitudes

Since the dissolution of the Soviet Union, local public and local government attitudes toward nuclear power have changed. Indeed, the ubiquitous NIMBYism of the period 1988–1990 has abated and in many places been replaced with a more grudging acceptance of industrial facilities, particularly those related to energy. An *Izvestiya* headline from September 1991 symbolized the change in attitude: "Are the Greens Always Right?"[23] The statements of Janos Tamulis, a member of parliment and prominent figure, about the bitterly opposed Ignalina station of the Lithuanian Green movement, serve as a striking anecdote:

> It is impossible to shut down [Ignalina] now because the electricity it produces serves Latvia, Belorussia and Kaliningrad. In 1988, that would have been Moscow's problem. Now it is a problem for Lithuania and other countries.[24]

Since 1991, in the Russian Federation there has been a fairly unambiguous and widespread support by local governments for nuclear projects. In some cases, these are clear reversals of past positions. Local authorities around the Balakovo, Kalinin and Kursk AES now support expansion of exsisting capacity at those stations.[25] Similarly, local officials in Vladivostok in the Russian Far East have supported the construction of a new nuclear power station; and in the Murmansk Oblast' authorities are supporting the expansion of exsisting capacity at the Kola AES.[26] Significantly, local authorities in the case of the Kola AES, have placed several conditions on the expansion project including free electricity and heat for workers at the station and increased investment in social provisions. Even in the newly independent state of Belarus (formerly the Belorussian SSR), which was considerably affected by the Chernobyl' accident and where antinuclear sentiment was strong, state officals have openly advocated the construction of nuclear plants as an alternative to the expensive importation of fuel.[27] However, such support for nuclear power is not ubiquitous, the planned Gor'kiy AST at Nizhniy Novgorod (Gor'kiy) was canceled by local oblast' and city authorities after years of controversy, even though it was ready for operation.[28] Support for nuclear power has not been unanimous; and in some cases there has been a clear divergence between local government and elements of the local public as in the case of the Kola AES in the Murmansk Oblast' where the local "green" movement continues to protest the station.[29]

There are several reasons for the apparent change in attitudes. Perhaps the most fundamental of these is the seriousness and magnitude of regional energy shortages throughout the former USSR. Both heating and particularly electricity shortages in conjunction with significant price increases in fuels have spurred local authorites and citizens to reassess the nuclear option for both heat and power. However, official positions for nuclear energy might reflect political maneuvering by local politicians to extract concessions from the Russian Federation in related energy issues. This appears to be the case in Armenia where local officials hope to guarantee the uninterrupted supply of fuels from Russia (a nuclear station in Armenia is a particularly unattractive prospect given the political and military conflict with neighboring Azerbaijan) and in Belarus where local officials hope to get lower fuel prices from Russia. Another possible explanation was recently indicated by Lord Marshall of Goring and Chairman of WANO who pointed out that nuclear power plants in the Russian Federation are provided with relatively high levels of social investment including ample supply of basic foods and facilities for their distribution, as well as educational and recreational facilities.[30] Admist the declining social and economic conditions in much of the Russian Federation, this must indeed be a powerful incentive for local government and even local residents to look favorably on nuclear projects.

The Third Dimension of Influence—The West

The previous discussion has illustrated that the unique conditions ongoing in the former Soviet Union appear to be evolving toward increased participation and influence in decision-making by local interests, as well as by national policy-makers and industry. Yet there is a third group of actors that in the last years of the USSR, and especially since the formation of the newly indepedent states, has been attempting to influence decision-making and policy—Western governments and transnational organizations. This is a unique condition of the complete lack of public and international confidence in the old Soviet nuclear power industry's handling of safety, and the opportunity for greater intervention as a result of the breakdown of the former economic and political system.

As pointed out earlier, increased local control in decision-making has been exercised by oblast' and city government. With the newfound sovereignty of the independent states of the former Soviet Union, policy-making and regulation in the nuclear industry has spatially devolved from the center to the former republics. Increasingly however, neighboring European countries are attempting to influence or intervene in policy-making, the formation of institutions and procedures, and even individual project making. Issues over which there is a battle for inflence by Western nations include: the continued operation of unsafe reactors (namely RBMKs and first-generation VVER-440s), evaluation and safety analysis of reactor technology, regulation of design and operations, and site evaluation.

Various nations within the structure of international organizations, such as the G-7 group of industrialized nations,[*] the European Economic Commission, the European Bank for Reconstruction and Development, and the World Bank, have attempted to compel the Russian Federation to discontinue operation of all RBMKs and first-generation VVER-440s in return for financial and other assistance in rebuilding and upgrading the Russian nuclear program.[31] Other foreign national and international organizations involved in efforts to improve design and operational

---

[*] These include the U.S., Japan, Germany, the U.K., France, Canada, and Italy.

safety and regulation in the former Soviet nuclear industry include
the U.S. Department of Energy, the U.S. Regulatory Commission,the
IAEA and WANO.[32]
    Several European and North American countries on a bilateral
basis are assisting the CIS countries in the evaluation and analysis
of reactor designs.   For example, the European Community is
assisting Russia in the evaluation of the RBMK design and Finland
is working with the VVER-91 reactor.[33] Similarly, several bilateral
programs are in progress to create and assist the regulatory
framework and monitoring for reactor design, siting and operation.
This is particularly important in Lithuania and Ukraine where the
withdrawal from the Soviet Union and Moscow-based regulatory
bureacracy has left these countries with an inadequate regulatory
infrastructure.   For example, Ukraine is receiving assistance from
France, and Lithuania from Sweden in forming regulatory codes,
licensing, and monitoring; Finland is helping improve management
and operations at both the Kola and Sosnovy Bor (formerly
Leningrad) plants.[34]
    Lastly, foreign experts are becoming involved in assisting local
authorities in site and technology decision-making.   In the St.
Petersburg Oblast' (formerly the Leningrad Oblast'), local officials of
the St. Petersburg Municipal and Regional Councils established a
competition for domestic and foreign plant designs of either nuclear
or nonnuclear technology to eventually replace the capacity at the
Sosnovy Bor AES. The evaluating jury consisted of twenty members,
including five nuclear experts from Western countries.*[35]

Making Sense of Soviet and Post-Soviet Experience

    The Soviet and post-Soviet experience with a high-risk
technology such as nuclear power, makes for some interesting
comparisons with what has been observed in the West. Clearly, up to
1986, the framework in which decisions were made concerning
location, scale, and technology in the Soviet nuclear power industry
can only be described as "elitist." A select group of decision-
makers—the national power utility *Minenergo*, national economic
planners, and the Communist Party leadership—maintained
supreme decision-making authority as well as a monopoly on
information. Influenced as they were by the research that indicated

---

* These Western members included representatives from Canada, France,
Germany, Sweden and the U.K.

that nuclear power could be a panacea to many of the Soviet Union's energy problems if nuclear reactors and plants were kept simple and built on a large scale, national economic planners and the party and state leadership embarked upon a massive nuclear power program. The political economy of the Soviet Union did not allow for participation by groups outside the traditional organs of economic planning whose interests were either directly or indirectly involved in a nuclear power project. If viewed from Kitschelt's framework of political opportunity structures, the Soviet Union, at least up to 1987–1988, maintained closed political structures for opponents of nuclear projects or programs. According to Kitschelt, other Western countries, France for example, shared similar closed avenues for opponents, albeit the level of control and intimidation by the leadership of the Communist Party and the Soviet state exceeded the French experience considerably.[36]

However, beginning in 1987 and escalating rapidly thereafter, opposition to nuclear power projects surfaced throughout the country taking on both assimilative and confrontational characteristics. The upsurge in opposition was triggered by the enormous loss in public confidence as a result of the Chernobyl' accident and by political reforms aimed, in part, at fostering greater participation among specialists and the public in decision-making and policy implementation throughout Soviet society. Ironically, these political reforms, while intended to expose the waste and corruption brought about by institutional elistism or "establishmentarianism", opened up discussion and avenues for political action that ultimately questioned the wisdom of party policies and legitimacy of existing institutions in the Soviet Union.

Some observers, Ziegler for example, have pointed out that environmentalism and the local opposition to industrial projects that sprang up during 1987–1991 had strong elements of nationalism and populism.[37] Indeed, even before the advent of *glasnost'*, the few cases of environmental opposition to industrial projects such as at Lake Baikal in the 1970s or the Siberian river diversions of the 1980s, contained some elements of nationalism and populism.[38] This is not suprising because the relative ineffectiveness of expert testimony against projects (ministries countered outside experts with thier own armies of experts armed with mountains of data) meant that appeals to nationalism were a relatively effective way to influence the party leadership against the project in question.[39] Although nationalism was certainly pervasive for many environmental issues and indeed in many movements against nuclear projects, particularly in the Ukrainian and Lithuanian Republics and the Tatar ASSR, not all antinuclear movements in the former USSR carried the specter of nationalism. Perhaps localism, best describes the underlying theme behind the antinuclear movement in the Soviet Union. These struggles essentially represented a desire by both local authorities and the local populace to gain some degree of control over projects that could have disasterous effects for those communities living around the nuclear plants in question. As with earlier environmental opposition in the Soviet Union, opponents of projects had to base their opposition on arguements consistant with established goals of the Party such as the preservation of unique national resources, reducing bureacratic inefficiency and waste, and providing for public safety.

In their struggles with the ministry officials and planners in Moscow, local opponents first used assimilative forms of expression (i.e., strategies by opponents working within established political processess and institutions) aimed at influencing national-level policy-makers in the nuclear issue included lobbying, petitioning, and political referenda. On the other hand, assimilative strategies aimed at affecting the implementation of exsisting policy, such as interventions in licensing hearings or litigation in the courts, were absent because of the lack of such procedural structures in the former Soviet Union. Local politicians and the public did, however, adopt confrontational strategies when a recalcitrant nuclear industry failed to respond to public demands (for example, the experience at the Crimean AES). Such confrontational strategies, in many cases reminiscent of the anti-nuclear movements in France, West Germany, and the United States, included public demonstrations, strikes, and plant blockades and, in some cases, were supported by local politicians and even local government.

Coincident with the dynamic changes affecting various forms of political expression, the ability of state institutions to implement policy eroded as well. It appears that by late 1990, there had been a significant devolution of authority and that some local political units had indeed gained some influence over existing Soviet state institutions, or if not influence over, then at least the ability to avoid implementing projects or directives contrary to their interests. While the outcome of ongoing struggles between state institutions and local political bodies at the republic, oblast', and city level had not yet played itself out, by the time of the August 1991 coup, it is clear that local political bodies had acquired the ability to frustrate and delay projects traditionally implemented by the central government.

However, since 1991 authorities in the newly independent states of the former Soviet Union as well as local authorities in the Russian Federation appear, in some cases, to have reversed their NIMBYish attitudes toward nuclear power. The political leaderships in the independent states of the Russian Federation, Lithuania, Kazakhstan, and even Armenia and Belorus (with no major research and development infrastructure or investment in this technology) have openly discussed the possibility of at least continuing if not expanding the nuclear option. In Ukraine, where public opposition to nuclear power seems strong, the executive leadership appears committed to nuclear power. The citizenry in many of these states appear to be quietly reassessing nuclear power in the face of a declining access to energy. Indeed it appears that the apparent NIMBYism during 1987–1991 was more a manifestation of center-periphery conflicts between Moscow and the localities, rather than a genuine antipathy toward nuclear power.

Despite the dissolution of the traditional Soviet forms of economic and industrial decision-making, it is by no means clear if more open, participatory institutions and procedures will emerge in the nuclear industries of CIS and non-CIS states of the former Soviet Union. Nevertheless, the experience of the Soviet nuclear power industry has left a profound legacy for these newly independent states. The poor record of Soviet planning and administration in nuclear power and the resulting public opposition to this has legitimized and strengthened the cause for local sovereignty and local participation in decision-making. Decentralization and increased local control over siting and project decisions does not, however, ensure enhanced safety. Energy shortages are common throughout most of the newly independent states, as the combination of insufficient financial resources, local opposition to energy projects, and the breakdown in intraregional trade in primary fuels has emaciated power production. National policy-makers in these new countries might indeed be forced to take desperate steps to stave off the looming energy crises. It remains to be seen how the local interests and the political and economic "powers that be" in the constituent states of the former Soviet Union will resolve the current regional economic and energy crises that prevail.

Notes

1. Matthew Sagers, "News Notes," *Post-Soviet Geography* (formerly *Soviet Geography*), Vol. 33, No. 4 (April 1992), p. 237. The last absolute decline in electric power production occurred at the height of the German invasion in 1942; Minenergo, *Razvitiye elektroenergeticheskogo*, p. 51.

2. "Russians Losing Hope That Nuclear Aid Will Materialize from West," *Nucleonics Week*, Vol. 33, No. 30 (July 23, 1992), pp. 10–13.

3. "Datafile: ex-USSR," *Nuclear Engineering International*, August 1992, p. 37

4. Ibid.

5. "Russians Losing Hope," pp. 10–11.

6. "IAEA Conference Repeats Promises but Produces No Safety Aid to East," *Nucleonics Week*, Vol. 33, No. 43 (October 22, 1992), pp. 7–8.

7. "Datafile: ex-USSR," p. 39.

8. "Local Authorities Will Make Decisions on Building Nuclear Power Plants," *CDPSP*, Vol. 44, No. 26 (July 29, 1992), p. 34.

9. "Russian Government is Very Short of Energy," *CDPSP*, Vol. 44, No. 22 (July 1, 1992), p. 22.

10. Ibid.

11. The announcement came in the form of Government statement No. 1026 (December 28, 1992) and was apparently drawn from a *Minatom* draft plan from the summer of 1992. "Russia Okays Plan to Proceed with Major Nuclear Construction," *Nucleonics Week*, Vol. 34, No. 3 (January 21, 1993), pp. 12–13.

12. Ibid.

13. Ibid.

14. "IAEA Conference Repeats," pp. 7–8.

15. "Ukrainian Lawmakers Striving to to Limit Nuclear Utility Power." *Nucleonics Week*, Vol. 33, No. 40 (October 8, 1992), pp. 12–13.

16. "Experts Say Human Error May Have Led to Leningrad-3 Failure," *Nucleonics Week*, Vol. 33, No. 18 (April 30, 1992), p. 11.

17. "Ukrainian Lawmakers," pp. 12–13.

18. "Experts Say Human Error," p. 11.

19. "Lithuanians See Possible Swedish Models for Fuel Storage, Outages," *Nucleonics Week*, Vol. 33, No. 36 (September 3, 1992), p. 12.

20. "Kazakhstan Plans New FBR at Shevchenko Research Complex," *Nucleonics Week*, Vol. 33, No. 44 (October 29, 1992), pp. 7–8.

21. Ibid.

22. "Armenia-2, Old VVER-440 is to be Readied for 1992 Restart," *Nucleonics Week*, Vol. 33, No. 2 (January 9, 1992), p. 4.

23. D. J. Peterson, "Hard Times for the Environment," *RFE/RL*, Research Institute, Report on the USSR, Vol. 3, No. 46 (November 15, 1991), p. 17.

24. Ibid.

25. "Russia Okays Plan," pp. 12–13.

26. Sagers, "News Notes," April 1992, p. 268.

27. Ibid.

28. Ibid.

29. Ibid.

30. "Understanding the CIS: Lord Marshall's View," *Nuclear Engineering International*, September 1992, p. 15.

31. "Lithuanians Face Myriad Problems at Ignalina Nuclear Station," *Nucleonics Week*, Vol. 33, No. 16 (April 16, 1992), pp. 7–8; "European Finance Heads Want More Eastern Reactor Safety Upgrades," *Nucleonics Week*, Vol. 33, No. 17 (April 23, 1992), pp. 1, 9.

32. "Datafile: ex-USSR," p. 36.

33. "Sidorenko Says Minatom-Financed Backfits of RBMKs Will Continue," *Nucleonics Week*, Vol. 33, No. 19 (May 7, 1992), p. 4; "Russians Favor VVER-1000 as Replacement Generation at Kola," *Nucleonics Week*, Vol. 33, No. 26 (June 25, 1992), p. 13.

34. "European Finance Heads," pp. 1, 9.

35. "Russians Open Plant Selection to International Jury," *Nucleonics Week*, Vol. 33, No. 12 (March 12, 1992), p. 1, 7; "Jury Gives Russia Choice of 11 Viable Reactor Designs," *Nucleonics Week*, Vol. 33, No. 23 (June 4, 1992), p. 13; "Datafile: ex-USSR," p. 36.

36. Kitschelt, pp. 74–75, 84–85.

37. Charles E. Ziegler, "Political Participation, Nationalism and Environmental Politics in the USSR," in John Massey Stewart (ed.), *The Soviet Environment: Problems, Policies and Politics* (New York: Cambridge University Press, 1992), pp. 24–35.

38. Donald R. Kelley, "Environmental Policy-Making in the USSR: The Role of Industrial and Environmental Interest Groups," Soviet Studies, Vol. 28, No. 4 (October 1976), 575–579; Robert G. Darst, "Environmentalism in the USSR: The Opposition to the River Diversion Projects," Soviet Economy, Vol. 4, No. 3 (July–September 1988), pp. 223–252.

39. Kelley, pp. 578–589.

Appendix A

Operational Reactors and Stations in the USSR

Station	Location	Reactor	Power capacity (MW)	Heating capacity (Gcal/h)	Reactor type	Construction start	Start-up date	Shutdown date
Ul'yanovsk AES	Dimitrovgrad Ul'yanovsk Ob. RSFSR	1	50		VK-50	NA	1965	
		2	12		BOR-60	NA	1969	
Beloyarsk AES	Zarechnyy Sverdlovsk Ob. RSFSR	1	100		AMB-100	NA	1964	1985
		2	160		AMB-200	1956	1967	1988
		3	600	280a	BN-600	1966	1980	
Novovoronezh AES	Novovoronezhskiy Voronezh Ob. RSFSR	1	210		VVER-210b	NA	1964	1987
		2	365		VVER-365b	1964	1969	1990
		3	417		VVER-440	1967	1971	
		4	417		VVER-440	1967	1972	
		5	1000	100a	VVER-1000	1974	1980	
Leningrad AES	Sosnovy Bor Leningrad Ob. RSFSR	1	1,000/700c		RBMK-1000b	1970	1973	
		2	1,000/700c		RBMK-1000b	1970	1975	
		3	1,000		RBMK-1000	1973	1979	
		4	1,000		RBMK-1000	1975	1981	
Shevchenko ATETs	Shevchenko Mangyshlak Ob. Kazakh SSR	1	150		BN-350	1967	1973	
Kola AES	Polyarnyye Zori Murmansk Ob. RSFSR	1	440	25	VVER-440b	1970	1973	
		2	440	25	VVER-440b	1973	1974	
		3	440	25	VVER-440	1977	1981	
		4	440	25	VVER-440	1976	1984	

Appendix A, Continued

Station	Location	Reactor	Power capacity (MW)	Heating capacity (Gcal/h)	Reactor type	Construction start	Start-up date	Shutdown date
Bilibino ATETs	Bilibino Chukotkaya AOk RSFSR	1	12	25	EGP-6	1970	1974	
		2	12	25	EGP-6	1970	1974	
		3	12	25	EGP-6	1970	1975	
		4	12	25	EGP-6	1970	1976	
Kursk AES	Kurchatov Kursk Ob. RSFSR	1	1,000/700c		RBMK-1000b	1972	1976	
		2	1,000/700c		RBMK-1000b	1973	1979	
		3	1,000		RBMK-1000	1978	1983	
		4	1,000	350a	RBMK-1000	1981	1985	
Armenian AES	Metsamor Armenian SSR	1	408	25	VVER-440b	1973	1976	1989
		2	408	25	VVER-440b	1975	1979	1989
Chernobyl' AES	Pripyat Kiev Ob. Ukrainian SSR	1	1,000/700c		RBMK-1000b	1972	1977	
		2	1,000/700c		RBMK-1000b	1973	1978	
		3	1,000		RBMK-1000	1977	1981	
		4	1,000		RBMK-1000	NA	1983	1986
Smolensk AES	Desnogorsk Smolensk Ob. RSFSR	1	1,000		RBMK-1000	1975	1982	
		2	1,000	200a	RBMK-1000	1976	1985	
		3	1,000		RBMK-1000	1984	1990	
Rovno AES	Kuznetsovsk Rovno Ob. Ukrainian SSR	1	402		VVER-440	1976	1980	
		2	416		VVER-440	1977	1981	
		3	1,000	100a	VVER-1000	1981	1986	
South Ukraine AES	Konstatinovka Ivano-Franovsk Ob. Ukrainian SSR	1	1,000		VVER-1000	1977	1982	
		2	1,000		VVER-1000	1979	1985	
		3	1,000	260a	VVER-1000	1985	1989	
Kalinin AES	Udomlya Kalinin Ob. RSFSR	1	1,000		VVER-1000	1977	1984	
		2	1,000	260a	VVER-1000	1982	1986	

Appendix A, Continued

Station	Location	Reactor	Power capacity (MW)	Heat ing capacity (Gcal/h)	Reactor type	Construction start	Start-up date	Shutdown date
Zaporozh'ye AES	Energodar Zaporozh'ye Ob. Ukrainian SSR	1	1,000	100	VVER-1000	1980	1984	
		2	1,000	100	VVER-1000	1981	1985	
		3	1,000	100	VVER-1000	1982	1986	
		4	1,000	100	VVER-1000	1984	1986	
		5	1,000	100	VVER-1000	1985	1989	
Ignalina AES	Snieckus Lithuanian SSR	1	1,500/1,250c		RBMK-1500	1977	1983	
		2	1,500/1,250c		RBMK-1500	1978	1987	
Balakovo AES	Balakovo Satatov Ob. RSFSR	1	1,000		VVER-1000	1980	1985	
		2	1,000		VVER-1000	1981	1987	
		3	1,000		VVER-1000	1982	1988	
Khmel'nitsky AES	Neteshin Khmel'nitskiy Ob. Ukrainian SSR	1	1,000		VVER-1000	1981	1987	

a Total capacity reported for station. Beloyarsk AES total probably lower, the figure given includes Beloyarsk-2 reactor that was subsequently shutdown.

b First-generation reactor of respective type.

c Power decreased to lower figure as mandated by *Gospromatomenergo* as of June 1990.

Notes: Location refers to nearest town, and the oblast' (Ob.) and republic in which station is located. Construction start refers to when work began on reactor foundation. Start-up date refers to year in which reactor was connected to the power grid for power supply.

Sources: IAEA, *Operating Experience*, 1988, pp. 461–549; IAEA, *Operating Experience*, 1991, pp.487–577; N. M. Sinev and B. B. Baturov, *Ekonomika atomnoy energetiki* (Moscow: Energoatomizdat, 1984), p. 51; Petros'yants, *Atomnaya nauka*, pp. 63–64; "Datafile: ex-USSR," p. 37.

Appendix B

Public Opposition at Selected Sites, 1986–1991

Station	Form of opposition	Major issues	Request for government commission (date)	Project canceled or suspended (date)	Project status at time of opposition
Krasnodar AES	Letters/Petitions	Seismicity	12/88	Canceled (1/88)	Site preparation
Chigirin AES	Letters	Ecology/Cultural		Canceled (5/89)	Site preparation
Odessa ATETs	Letters/Petitions	Seismicity/Proximity		Canceled (9/88)	Under construction
Kharkov ATETs	Letters	Proximity		Canceled (9/88)	Under construction
Minsk ATETs	Letters/Petitions	Proximity		Canceled (9/88)	Under construction
Crimean AES	Letters/Petitions/Strikes	Seismicity	12/88	Canceled (10/89)	Under construction
South Ukrainian AES	Letters/Petitions	Ecology/Proximity	12/88	3rd Reactor canceled (12/89)	Operating, expansion
Ignalina AES	Letters/Petitions/Protests Blockades	Ecology	9/88a 6/89b	3rd, 4th Reactors canceled (9/90)	Operating, expansion
Armenian AES	Letters	Reactor Safety Seismicity		Shutdown (3/89)	Operating
Leningrad AES	Letters/Petitions	Reactor Safety			Operating
Far East AES	Letters/Petitions	Ecology/Seismicity	1/89	Site investigation ongoing	Site preparation
Gor'kiy AST	Letters/Petitions	Proximity	3/89	Canceled (12/90)	Under construction
Yaroslavl' ATETs	Letters	Ecology	5/89	Canceled (4/90)	Site preparation
Tatar AES	Letters/Protests/Referendum	Ecology/Seismicity Proximity	1/89		Under construction
Bashkir AES	Letters/Protests/Referendumc Strike/Blockade	Ecology/Seismicity	10/89		Under construction
Rostov AES	Letters/Petitions/Protests Blockade	Ecology Geology Proximity	7/89	Canceled (9/90)	Under construction
South Urals AES	Letters	Ecology			Under construction
Karelian AES	Letters/Petitions/Protests	Ecology			Site preparation
Voronezh AST	Letters/Petitions/Referendum	Proximity		Canceled (12/90)	Site preparation
Arkhangel'sk AST	Letters/Petitions	Geology/Proximity		Canceled (2/90)	Under construction
Balakovo AES	Letters/Protests	Ecology	6/90	5th, 6th Reactors	Operating, expansion

Appendix B, Continued

Station	Form of opposition	Major issues	Request for government commission (date)	Project canceled or suspended (date)	Project status at time of opposition
Khmel'nitskiy AES	Letters/Protests/Blockade				Operating, expansion
Chernobyl' AES	Letters/Protests	Reactor Safety			Operating, expansion
Zaporozh'ye AES	Letters/Petitions/Protests	Ecology		6th Reactor canceled (12/90)	Operating, expansion
Kursk AES	Letters/Petitions			6th Reactor canceled (4/89)	Operating, expansion
Kostroma AES	Letters	Ecology/Geology			Under construction
Azerbaijan AES	Letters	Seismicity		Canceled (1/89)	Site preparation
Georgian AES	Letters	Seismicity		Canceled (1/89)	Site preparation
Estonian AES	Letters	Reactor Safety			Site preparation
Rovno AES	Letters	Reactor Safety			Operating, expansion
Belorussian AES	Letters			Canceled (12/88)	Site preparation
Volga AES	Letters			Canceled (4/90)	Site preparation
Perm AES	Letters				Site preparation
Latvian AES	Letters	Ecology		Canceled (8/89)	Site preparation

[a] Request for the investigation of Ignalina 3.

[b] Request for the investigation of the entire station.

[c] Referendum not recognized by local and All-Union authorities.

Notes: Major issues include those site-related issues that project opponents cited in their opposition to the project. Request for Expert Committee includes official requests to All-Union or republican authorities for a government commission. Project cancellation refers to official cancellation by the USSR Council of Ministers.

Sources: Compiled by the author from various articles in *FBIS, CDSP, RFE/RL, Izvestiya, Pravda, Sotsialicheskaya Industriya, Sovetskaya Rossiya,* and *Soviet Geography.*

Appendix C

Public Opposition to Selected Power Stations

For the convenience of the reader, the following appendix is provided presenting, on a site-by-site basis, a description of public and local government opposition to 25 of the most controversial stations during the period 1986–1991.

Krasnodar AES

An early site that met public opposition was the Krasnodar AES in the Krasnodar Kray in the North Caucasus. Public opposition to the Krasnodar AES in the form of letters and petitions to government officials and the media continued through the summer and fall of 1987 until the station was canceled in January 1988.[1] The main concern of the local populace to the Krasnodar AES was plant safety as well as site selection, in particular the station's location in a known seismically active area.[2]

Chigirin AES

Another early episode of public opposition to a nuclear project was the Chigirin AES in the Ukrainian SSR. The opposition to the Chigirin AES station was given considerable publicity. The site of this planned station was on the banks of the Dnepr Reservoir in the Poltava Oblast' in the central Ukraine. A letter openly opposing the site of the proposed Chigirin AES was submitted to the editors of *Literaturna Ukraina* and published on August 6, 1987.[3] This letter, submitted and signed by cultural figures (writers and historians) and even local Party officials (including the First Secretary of the Poltava Oblast' Party Committee) criticized the proposed station's location for several reasons: the station would contribute to an already excessive level of water withdrawals from the Dnepr River; local residents had not been consulted; and the site is a well-known historical landmark in Ukraine, being the former capital of the 17th-century Ukraine.[4]

By 1988 several letters by Ukrainian scientists and ecological groups were published in the Ukrainian media. Among these were two prominent letters from Ukrainian scientists who objected to Chigirin outright, as well as opposing capacity expansions at three exsisting stations—Rovno, Khmel'nitskiy, and South Ukraine AESs.[5] In December 1988, another letter submitted by *Zelenyi Svit* (Ukrainian for Green World) was published in *Vechirnii Kyiv*.[6] This letter demanded the immediate cessation of construction at the Crimean, Chigirin, and South Ukraine AESs, pending a judgment by a body of experts on the relative safety of the site, station design, and construction. Similar demands were presented about the Crimean and Chigirin stations in the draft program of the Ukrainian Popular Front.[7] After continued protest through the Ukrainian media, the Ukrainian State Planning Committee and the USSR Council of Ministers announced on May 19, 1989, that the plans for the Chigirin AES had been canceled.[8]

Odessa ATETs

Another station that was the subject of local opposition in Ukraine during the fall of 1987 and spring of 1988 was the Odessa ATETs. The Odessa ATETs was located in the Odessa Oblast' in the southwestern Ukraine. The ATETs designs required this type of nuclear station to be within 40 km of consuming centers. In the case of Odessa, the Odessa ATETs was located 25 km (15.5 miles) from the city. As a result, public concern in and around Odessa was considerable. Ukrainian scientists criticized the Odessa ATETs both because of its proxmity to such a sizable city (in 1989 Odessa's population was 1,115,000 inhabitants) and its location on an earthquake fault line.[9] Public opposition was apparently a major factor in the reassessment of the ATETs design by Soviet power engineers and energy policy-makers. In September 1988, it was announced that the Odessa ATETs as well as the Khar'kov ATETs and Minsk ATETs (all similar designs) were canceled.[10]

Khar'kov ATETs

The Khar'kov ATETs was located just outside Khar'kov in the Khar'kov Oblast' in the eastern Ukraine. Like the Odessa ATETs it was a cogeneration design planned to be in close proximity to the city of Khar'kov. There was local public oppostion to the Kharkov station and expert discussion on the merits of the ATETs design and the station's location during the spring of 1988.[11] In February 1988, it was reported in the Ukrainian press that the Khar'kov ATETs was to be resited, apparently in response to public concerns over its proximity to the city of Khar'kov. Nevertheless, public opposition continued. The Khar'kov ATETs was canceled in September 1988 in conjunction with several other ATETs projects.[12]

Minsk ATETs

The Minsk ATETs was planned to be built approximately 37 km from the city of Minsk in Vitebsk Oblast' in the central Belorussian SSR. The second of a series of ATETs planned for large urban centers, the Minsk ATETs was to become operational by 1989.[13] Opposition to the station began in earnest in 1988. In September 1988, after Belorussian authorities, including the Belorussian Party and Council of Ministers, requested a reevaluation of the project by All-Union authorities, it was announced that the Minsk ATETs was to be canceled and was being replaced by a gas-fired TETs.[14]

Crimean AES

During 1988, a fifth Ukrainian site, the site for the Crimean AES, became a focus of public attention and debate. The Crimean AES was located in the Crimea Oblast' in southern Ukraine. The actual site was located on the coast of the Sea of Azov on the Kerch Peninsula in the Crimea. Open opposition to the Crimean AES surfaced in May 1988, when the Kiev paper *Kul'tura i zhyitya* published an open letter from seven cultural figures from Sevastopol' criticizing plans for the Crimean station. The letter was accompanied by the signatures of 475 residents of Sevastopol'.[15] In December 1988, *Zelenyi svit* included the Crimean AES among the nuclear projects that should be reassessed for siting and design reasons.[16] Criticisms of the Crimean AES even extended outside of Ukraine. Students at Moscow State University signed their own petition against the Crimean site.[17] The primary concerns over the Crimean AES were largely site related. Opponents of the project argued that the area was seismically active and that a nuclear station would endanger the pristine coastline of a popular rest and vacation area.[18]

In response to public concerns and an appeal from the Ukrainian Communist Party Central Committee, the USSR Council of Ministers established a special Government Commission to reexamine the plans for the Crimean AES.[19] This special commission (apparently the first of its kind in the USSR), chaired by

Ye. Velikov, Vice President of the USSR Academy of Sciences, was organized into four specialized working groups to investigate site-related safety issues, particularly seismicity. The commission's final decision was to be discussed and approved at a session of the AN SSSR Presidium's Interdepartmental Council on Seismology and Earthquake-Resistant Construction (MSSSS)*. The commission's working groups found that there was considerable seismic risk at the site of the Crimean AES. Geologic evidence indicated that Force 9 earthquakes had occurred at the site of the Crimean AES.[20] Moreover, the commission found that in the immediate vicinity of the station was a subterranean dome. Gases under high pressures within this dome could lead to deformations of the earth's surface and to mud or gas volcanism. The MSSSS in September 1988 supported the conclusions of the Government Commission concluding "their [seismic events'] direct influence on the AES structures could exceed the anticipated seismic influences in terms of destructive consequences."[21] The special Government Commission and the MSSSS final recommendation accepted the estimate of a Force 9 earthquake as the "maximum potential earthquake" for the Crimean site and added that there was a danger of mud volcanism at the site. These findings were bitterly attacked by specialists within the *Minatomenergo*.[22] Although the findings of the Government Commission were given nationwide coverage in the media, the final decision concerning the Crimean AES was left to the USSR Council of Ministers. Almost a year after the special Government Commission made its final report, the USSR Council of Ministers, on October 29, 1989, decided against completing the Crimean AES.[23]

South Ukraine AES

While the planned stations were the focus of public opposition based on site-related concerns, existing nuclear stations soon became embroiled in public debate over similar issues. Opposition to existing stations, like opposition to planned stations, was due to a lack of confidence in the advisability of their location, design, and potential environmental impact. One such operational station that became the object of local opposition was the South Ukraine AES. The South Ukraine AES was located in the Nikolayev Oblast' in southern Ukraine. At the 19th Party Conference in June 1988, Ukrainian

---

* MSSSS - Mezduvedomstvennogo soveta po seysmologii i seysmostoykomu stroitel'stvu

scientists and cultural figures included the South Ukraine AES among the criticized projects.[24] In December of that same year, *Zelenyi svit* criticized the South Ukraine AES, along with the Crimean and Chigirin AES sites, demanding an immediate cessation of construction of a third reactor until an outside body of experts could be found to decide whether the site and station were safe.[25] In April 1989, a petition signed by more than 200,000 residents of Nikolayev requested a halt to the construction at the South Ukraine AES.[26] In December 1989, it was reported that the third reactor at the South Ukraine AES was canceled by the USSR Council of Ministers.[27]

Ignalina AES

In the Lithuanian SSR, a protest movement soon coalesced around the Ignalina AES, located in the southeast corner of the republic. Between 1986 and 1988, a few Lithuanian scientists had shown concern about the radiation and thermal effects of discharged cooling water on the local environment.[28] Public protests against the Ignalina AES followed in the summer of 1988; the majority of these early public protests focused on the scale and technology of the plant (the reactor was an RBMK, the same basic design as the reactors at Chernobyl'). Local criticism of the plant also focused on the fact that the reactor buildings were known to have settled into the ground.[29] Before the Chernobyl' accident, the Ignalina AES was planned to eventually have four RBMK-1500 reactors, with a gross capacity of 6,000 MW. In April 1988, the planned fourth reactor was canceled. According to Lithuanian authorities, this was due to major planning oversight—there was inadequate cooling at the site for all four RBMK-1500 reactors.

Nevertheless, during the summer of 1988, Lithuanian experts and the public opponents of the Ignalina AES focused their attention on the planned third reactor. For example, in June the Lithuanian paper *Kojaunimo Tiesa* published an article by the deputy chairman of the ecology group *Zemyna* criticizing the plans for the third reactor because of the damaging thermal pollution it would incur on Lake Druksiai (used as a cooling reservoir). The article also advocated the installation of a smaller capacity VVER-1000 rather than the RBMK-1500 if a third reactor was to be built.[30] A minor electrical cable fire in September further stimulated public resistance to plans for the third reactor. On September 10, 1988, the Lithuanian Movement for Perestroika called for a blockade of the plant by protesters.[31] By mid-September 1988, the Initiative Group for the Support of Perestroika submitted a petition signed by more than 287,000 individuals requesting that the plant be investigated by an international commission.[32] In response to this public pressure, the Lithuanian Communist Party formed an independent commission to investigate the need for a third reactor.[33]

The Ignalina AES is under the Ministry of Energy of the independent Lithuanian government. As of December 1992, it appears that a third reactor is not planned for the Ignalina station; however, the Lithuanian Ministry of Energy is interested in eventually replacing Ignalina-1 and -2, with an upgraded, safer and probably western reactor design

Belorussian AES

In the Belorussian SSR, public opposition to plans for a nuclear power station surfaced in the fall of 1988. The November 2, 1988, issue of the All-Union weekly *Literaturnaya gazeta* included a letter by seven Belorussian cultural figures and scientists emphasizing the local public opposition to plans for a nuclear power station in the Vitebsk Oblast' (referred to as the Belorussian AES).[34] Opposition to the plant was based on concerns over ecological damage. Apparently, the Belorussian AES was still in the initial planning and surveying stage. A month later on December 13, 1988, Belorussian authorities announced that all plans for the Belorussian AES had been canceled.[35]

Armenian AES

The Armenian earthquake of December 7, 1988, brought attention to the Armenian AES west of Yerevan in the center of the Armenian SSR. During the winter of 1988–1989, the Armenian earthquake played the role of a catalyst for public opposition to nuclear power projects and further undermined public confidence in official decision-making. All-Union authorities quickly decided to permanently shut down both reactors at the Armenian AES by 1991 (eventually Armenian-1 was decommissioned on February 25, 1989, and Armenian-2 on March 18, 1989).[36] Nevertheless, in the course of public discussion on the station's safety in the aftermath of the earthquake, there were relevations that called into question the competence of established decision-making procedures and processes. According to Yevgeniy Velikhov, Vice President of the USSR Academy of Sciences, the issue of shutting down the

Armenian AES because of seismic concerns surfaced in 1985 after a seismic zoning map was compiled by experts who indicated that a Force 8 earthquake was possible at the site of the station. As a result of this survey, taken some 14 years after the Armenian AES had gone into operation, the station was required to enhance the earthquake resistance of its buildings and equipment.[37] At the time of the December 7, 1988, earthquake, the required improvements had not been completed.[38] Although the Armenian AES survived the December 7 earthquake undamaged (the earthquake registered Force 5.5), the failure of the initial site selection process to identify adequately the potential seismic risk at the site and the failure of ministry authorities to improve building resistance in a timely manner called into question, in the public's eye, the competence of power ministry officials. The Armenian AES was shut down in 1989.

Far East AES

Plans for the Far East AES in the Khabarovsk Kray in the Soviet Far East attracted public protest during the winter of 1988– 1989. Petitions against the projected station, located north of Komsomol'sk-na-Amure, were submitted to local authorities by inhabitants in Khabarovsk, Komsomol'sk-na-Amure, and Amursk.[39] Public concerns centered on site selection and station construction. In response, local Communist Party officials organized a public meeting (televised locally) between representatives of the Ministry of Nuclear Power and Nuclear Industry and the local population. As a result of the public concern, the Khabarovsk Kray's Krayispolkom established a commission of experts to investigate site selection for the Far East AES.[40] The search for a Far East AES was still ongoing in December 1992.

Gor'kiy AST

During the spring and summer of 1989, public opposition intensified against stations along the Volga River. One of the stations planned along the Volga was in the immediate vicinity of Gor'kiy (renamed Nizhniy Novgorod in 1989, a city of 1,438,000 inhabitants) in the Gor'kiy Oblast'. Plans for a nuclear station at Gor'kiy included the installation of the prototype and highly controversial AST design (see Chapter 3 for discussion on the AST design). The Gor'kiy AST, the first of its kind in the USSR or the world, was to be located some 12 km from the Gor'kiy city center and

5 km from a residential district.[41] Although letters of protest had been submitted during the preceeding two years, concerted public action did not surface until 1989. In March 1989, three scientists openly questioned the safety and neccessity of the Gor'kiy AST, arguing that gas-fueled boiler units would be safer and more economical. The authors also criticized the current decision-making procedure and advocated independent (i.e., nonministerial) scientific oversight including specialists from different academic disciplines, as well as public participation.[42] Despite a clean bill of health from a visiting group of IAEA experts in June, public opposition intensified.[43] In July 1989, more than 100,000 residents of Gor'kiy signed a petition demanding the cancellation of the Gor'kiy AST.[44] Although the Gor'kiy AST was ready for start-up in September 1989, the decision to do so was postponed as Gor'kiy city and Gor'kiy Oblast' authorites refused to allow its operation.[45]

Yaroslavl' ATETs

The spring and summer of 1989 also saw an intensification of public opposition elsewhere along the Volga. The Odessa, Khar'kov, and Minsk ATETs had been confirmed as canceled in September 1988. Nevertheless, plans for another ATETs for the city of Yaroslavl' had been disclosed in August 1988. Letters opposing the station were submitted by individual citizens as well as by several local Communist Party organizations, including labor collectives, trade union organizations, and the Communist Youth League.[46] In May 1989, *Sovetskaya Rossiya*, a traditionally conservative newspaper, published a lengthy article criticizing plans for the Yaroslavl' ATETs on safety and economic grounds.[47] The article stressed the neccessity of an independent expert commission to look into the Yaroslavl' project. On April 1, 1990, the project appeared to have been canceled, with a report that the Oblast' Party leadership had decided to abandon further planning work for the ATETs, opting instead for a gas-fueled generating station.[48]

Tatar AES

Another nuclear power station site on the Volga that received considerable attention in 1989–1990 was the Tatar AES, located in the Tatar ASSR. The Tatar AES, as mentioned in Chapter VI, was a focus of concern for some members of the local population even before the Chernobyl' accident. However, overt public protests against the station began in 1988.[49] In January 1989, several academics from Kazan University criticized the choice of the site, reactor safety, and neccessity of the Tatar AES.[50] Two demonstrations against the station followed at Nizhnekamsk and Kamskiye Polyany in April and July, respectively.[51] In July 1989, Gumer Usmanov, Secretary of the Tatar ASSR Regional Party Committee, and Mintimer Shaymiyev, Chairman of the Tatar ASSR Council of Ministers, formally requested that the subject be investigated in a comprehensive examination by a government commission.[52] A three-day march followed in early October, in which protesters marched from Kazan University to the station site at Kamskiye Polyany, a distance of approximately 100 km.[53] At the time of the march, scientists from the Central Scientific Research and Design Institute of Urban Planning and the All-Union Scientific Research Institute of Geology and Nonmetalliferous Minerals as well as scientists from the USSR Academy of Sciences and Kazan University published a letter in the local press.[54] In their collective opinion, the site of the Tatar AES was unsafe, they cited the large size of the station (a planned capacity of 8,000 MW) and subsequent massive water withdrawals in an already heavily industrialized area, and its location on a geologically active fault (an earthquake measuring Force 6 had occurred at Yelabuga, about 30 km north of Nizhnekamsk, on September 16, 1989). Protest rallies and public opposition by scientific experts continued through the spring and summer of 1990.[55] As of December 1991, the status of the Tatar AES was unresolved. Current plans of the Russian Federation's *Minatom* do not include this project.

Bashkir AES

In the adjacent Bashkir ASSR, public opposition against the planned Bashkir AES began 1988.[56] By 1990 public opposition had intensified considerably. A referendum held in Neftekamsk on March 1, 1990, indicated massive public dissatisfaction with the project, with 99% of the votes cast against the AES.[57] Public opposition became even more militant with a 24-hour blockade of the construction site on August 30, 1990 and a general strike in Neftekamsk on September 10, 1990.[58] After lengthy public debate, republic legislators (i.e., the Bashkir Supreme Soviet) passed a resolution against the project.[59] As of December 1992, the Bashkir AES appeared to have been cancelled.

Rostov AES

Another station that was the object of opposition was the Rostov AES, located along the Don River near Volgodonsk in the Rostov Oblast' in the southern RSFSR. Public opposition to the Rostov AES began in earnest during the summer of 1989. In June of that year, local authorities received a petition against the station with 58,000 signatures from the residents of Volgodonsk (a city of 176,000 people).[60] This petition, in particular, requested an investigation into the construction of the Rostov AES through a "non-departmental" independent commission. The main concern of project opponents was the unstable geology of the site and its close proximity to the city of Volgadonsk (the site was 13 km from Volgodonsk). Public demonstrations against the project continued into August, when protesters blockaded the Rostov AES and prevented construction and building supplies from being delivered to the site.[61] In February 1990, work on the Rostov AES was postponed by the USSR Council of Ministers until an independent commission made an assessment of the site.[62] Throughout the spring and summer of 1990, local officials and legislative bodies continued to oppose the project. In September 1990, construction of the Rostov AES was finally halted by the USSR Council of Ministers.[63]

South Urals AES

Opposition to the South Urals AES appears to have first surfaced in 1989. The South Urals AES is located in the Chelyabinsk Oblast' in the southern foothills of the Ural Mountains. Although this station was in an advanced stage of construction, an independent expert commission drawn from the USSR State Committee for the Protection of Nature reviewed the power station draft plans. The committee accepted the draft plans. However, a decision was made to not resume construction of the South Urals AES without public acceptance of the station from the inhabitants of the region.[64] Notably, local Oblast' officials including the legislature publically supported the project. Although public opposition has continued, the South Urals AES was reported to still be under construction as of August 1992 and was included in the Russian Federation's long-term plans for the nuclear power industry.[65]

Karelian AES

In March 1989, Soviet energy officials announced that a site was being selected for a base-load nuclear power station in the Karelian ASSR.[66] By January 1990, a site had apparently been selected in the Muyezersk Rayon. Public protests and petition drives opposing the station were quickly organized by the Student League of Greens and the popular front of Karelia.[67] In March 1990, after many rallies, petitions, and articles in the local press, the Karelian ASSR Supreme Soviet ordered a suspension of work on the planned station.[68] As of December 1991, the status of the Karelian AES was unclear with no official word from All-Union authorities.

Voronezh AST

The Voronezh AST was the object of considerable public opposition after the Chernobyl' accident. The Voronezh AST had been under construction since the early 1980s and was to be located less than 15 km from city of Voronezh (with a population of 887,000 in 1989).[69] The city of Voronezh conducted a referendum on May 15, 1990. More than 95% of those who participated voted against the project—some 500,000 city residents.[70] As a result of this referendum, construction of the station was canceled.

Arkhangel'sk AST

The plans to construct the Arkhangel'sk AST were first announced in 1987.[71] The site was just outside Arkhangel'sk in the Arkhangel'sk Oblast' in northern Russia. Specialists and the local public criticized the project in 1989 and 1990. Major criticisms of this project stemmed from the site's location next to a large subsurface reservoir of water as well as the fact that it was 5 km from the city of Arkhangel'sk.[72] In February 1990, the project was reported to be suspended.[73]

Balakovo AES

Public opposition arose to the continued expansion of the Balakovo AES in the Saratov Oblast' of the RSFSR. In June 1990, several rallies and protests opposed the addition of a fifth and sixth reactor at the Balakovo site.[74] Organized by the local Greens'

Movement, these protests also demanded independent expert examination of the project.[75] In November 1990, it was reported that construction had ceased on the fourth, fifth, and sixth units of the Balakovo AES as ordered by central authorities.[76] However in December 1992, the government of the Russian Federation announced that the fourth reactor at the Balakovo station would go into operation by the end of 1993.[77]

Khmel'nitskiy AES

Another Ukrainian power station that became embroiled in controversy was the Khmel'nitskiy AES. The Khmel'nitskiy AES was located in the Khmel'nitskiy Oblast' in western Ukraine. The first reactor went into operation after the Chernobyl' accident in 1987 and another reactor was scheduled to go into operation by 1990.[78] Although protests against expansion of the Khmel'nitskiy AES date from 1988, they intensified during the summer of 1990. Public opposition to the continued expansion of the Khmel'nitskiy AES took the form of rallies, hunger strikes, and supply interruptions by local enterprises who were providing materials to the project.[79] In July, protesters organized by Ukrainian activist organizations *Rukh* and *Zeleniy Svit* blockaded the Khmel'nitskiy AES in response to the oblast's unsuccessful attempt to pass a moratorium on further construction.[80] However, it appears that by August 1992 construction was still proceeding on the second reactor at the Khmel'nitskiy AES.[81]

Zaporozh'ye AES

Protests and petitions against the Zaporozh'ye AES in the Nikopol' Oblast' erupted during the summer of 1990. At that time five reactors were in operation at the site. Protests organized by Greenpeace and the local politicians from the Congress of People's Deputies demanded that the USSR Supreme Soviet establish an independent expert commission to reevaluate the site and determine the appropriateness of a sixth reactor that was under construction.[82] In December 1990, the USSR Council of Ministers announced that construction on the sixth reactor at the Zaporozh'ye AES had been suspended.[83] By the summer of 1992, construction had apparently resumed on the Zaporozh'ye-6 reactor.[84]

Kursk AES

At the Kursk AES, local residents had expressed opposition to the station since 1987. This opposition focused on deviations from construction plans during the building of the first two reactors before the Chernobyl' accident.[85] Residents also were opposed to the construction of the planned fifth and sixth units in 1990.[86] Although in September 1988, it was announced that the sixth reactor at Kursk would not be completed, as of December 1990 the status of the fifth reactor was unclear.[87] However, the new nuclear program of the Russian Federation announced in December 1992, includes the Kursk-5 reactor to be completed by 1995.[88]

Kostroma AES

The planned Kostroma AES was yet another victim of public opposition. The station was located in the Kostroma Oblast' in the RSFSR. In the case of the Kostroma AES, inadequate site analysis relating to local geology and ecology of the immediate area was the main issue of concern. Local political figures, state institutions, and public interest organizations orchestrated opposition to the station, including Oblast' People's Deputies, the department of the State Committee for the Protection of Nature in the Kostroma Oblast', and the Committee for the Protection of the Volga.[89] In 1990, the Kostroma Oblast' Soviet of People's Deputies voted to halt construction on the station, requesting the USSR Council of Ministers to heed its decision.[90] The Kostroma AES has not appeared in the subsequent plans of the Russia Federation's *Minatom*.

Notes

1. *Pravda*, January 21, 1988, p. 6.

2. Ibid.

3. See Chapter VI, p. 133, n. 24.

4. Marples, "Chigirin," pp. 26–29; Solchanyk, "Ukrainian Writers Protest ," pp. 1–2.

5. Nahaylo, pp. 1–2.

6. Marples, "Chigirin," pp. 26–29.

7. Ibid.

8. Ibid.

9. "Tatar Anti-Nuclear Protest Reported," p. 80.

10. *Izvestiya*, September 7, 1988, p. 2.

11. "Differing Views on Ukrainian Stations," *FBIS*, FBIS-SOV-88 032 (February 18, 1988), p. 88; "Precautions in Expanding Nuclear Power Viewed," *FBIS*, FBIS-SOV-88-046 (March 9, 1988), p. 47.

12. *Izvestiya*, September 7, 1988, p. 2.

13. "Atomic Power: Is the Impetus Being Lost?" *CDSP*, Vol. 40, No. 36 (October 5, 1988), pp. 8–9; David Marples, "New Protests against Soviet Nuclear Energy Program," *RFE/RL, Radio Liberty Research Reports*, RL 448/88 (September 28, 1988), p. 4.

14. "Atomic Power: Is the Impetus Being Lost?" pp. 8–9; Marples, "New Protests," p. 4.

15. Solchanyk, "Ukrainians Send Appeal," p. 4.

16. Marples, "Chigirin," p. 27.

17. Solchanyk, "Ukrainians Send Appeal," p. 4.

18. Ibid.; *Pravda*, January 11, 1989, p. 3.

19. *Pravda ukrainy*, January 15, 1989, pp. 1, 3.

20. *Pravda*, January 11, 1989, p. 3.

21. Ibid.

22. Ibid.

23. *Izvestiya*, October 27, 1989, p. 3.

24. *Literaturna ukraina*, June 23, 1988, p. 4.

25. Marples, "Chigirin," p. 27.

26. "Opposition to Nuclear Power Stations in the Ukraine," *BBC, SWB*, SU/0451 (May 5, 1989), C/3.

27. Marples, "Chigirin," p. 27.

28. Girnius, "The Ignalina Atomic Plant's Second Reactor," pp. 12–13.

29. Girnius, "Continued Controversy ," p. 30.

30. "Claim of Reactor Subsidence Denied," *FBIS*, FBIS-SOV-88-114 (June 14, 1988), p. 34.

31. "Group Calls for Blockade of Plant," *FBIS*, FBIS-SOV-88-176 (September 14, 1988), p. 56.

32. David Marples, "New Protests," p. 4.

33. Ibid.

34. *Literaturnaya gazeta*, November 2, 1988, p. 1.

35. "Belorussian Public Rejects Planned Nuclear Plant," *FBIS*, FBIS-SOV-88-239 (December 13, 1988), p. 63.

36. "USSR to Close Armenian Nuclear Plant to Meet Public Demands," *Nucleonics Week*, Vol. 29, No. 52 (December 15, 1988), p. 1.

37. "Armenian Nuclear Power Station Safety Examined," *FBIS*, FBIS-SOV-88-250 (December 29, 1988), p. 25.

38. Ibid.

39. *Pravda*, February 1, 1989, pp. 2–3.

40. Ibid.

41. "100,000 Sign Nuclear Power Plant Protest Message," *FBIS*, FBIS-SOV-89-146 (August 1, 1989), p. 81.

42. *Sotsialisticheskaya industriya*, March 26, 1989, p. 3.

43. Ibid.

44. "100,000 Sign," p. 81.

45. "Gor'kiy Soviet to End Power Station Project," *FBIS*, FBIS-SOV-90-104 (May 12, 1990), p. 83; *Izvestiya*, August 25, 1990, p. 2.

46. *Sovetskaya rossiya*, April 1, 1990, p. 1.

47. *Sovetskaya rossiya*, May 21, 1990, p. 3.

48. "Construction of Most Nuclear Plants Suspended," *FBIS*, FBIS-SOV-90-237 (December 10, 1990), p. 81.

49. Sheehy and Voronitsyn, p. 3.

50. "Officials Challenged," *BBC, SWB,* C2/1.

51. *Sovetskaya rossiya*, April 26, 1989, p. 3; "Tatar Anti-Nuclear Protests Reported," p. 79.

52. "Tatar Anti-Nuclear Protests Reported," p. 79.

53. *Pravda*, October 5, 1989, p. 6.

54. Ibid.

55. "Rally Protests Tatar AES Construction," *FBIS*, FBIS-SOV-90 123 (June 26, 1990), p. 105.

56. "Construction of Bashkiriya AES Halted," p. 106.

57. "Neftekamsk 'Referendum' Criticizes Bashkir AES," *FBIS*, FBIS-SOV-90-079 (April 24, 1990), p. 39.

58. "Bashkiriya Supreme Soviet Debates Nuclear Plant," *FBIS*, FBIS-SOV-90-172 (September 5, 1990), p. 117.

59. "Construction of Bashkiriya AES Halted," p. 106.

60. "From the Stenograph of the Congress of People's Deputies," *FBIS*, FBIS-SOV-89-142-S (July 26, 1989), p. 21; Population data from the Soviet 1989 Census as reported in Matthew Sagers, "News Notes," *Soviet Geography*, Vol. 30, No. 5 (May 1989), p. 425.

61. "Government Postpones Rostov AES Commissioning," *FBIS*, FBIS-SOV-90-039 (February 27, 1990), p. 93.

62. Ibid.

63. "Workers Protest Lost Jobs at Rostov Nuclear Plant," p. 45.

64. "Nuclear Power Station Causes Controversy," *FBIS*, FBIS-SOV-89-194 (October 10, 1989), p. 91.

65. "Chelyabinsk Approves Atomic Power Facility," *FBIS*, FBIS-SOV-90-227 (November 26, 1990), p. 59; "Chelyabinsk to Hold Nuclear Power Referendum," *FBIS*, FBIS-SOV-90-250 (December 28, 1990), p. 116–118; "Datafile: ex-USSR," p. 38; "Russia Okays Plans," p. 12.

66. "Chernobyl' Aftermath," p. 6.

67. "Protest against Karelian Power Plant Construction," *FBIS*, FBIS-SOV-90-012 (January 18, 1990), p. 132.

68. "Karelian Nuclear Project Suspended," *FBIS*, FBIS-SOV-90-042 (March 2, 1990), p. 84.

69. Population data from the Soviet 1989 Census as reported in Sagers, "News Notes," (May 1989), p. 425.

70. *Komsomol'skaya pravda*, May 24, 1990, p. 2.

71. Matthew Sagers, "News Notes," *Soviet Geography*, Vol. 28, No. 12 (December 1987), p. 778.

72. "Work Halted on Arkhangel'sk," pp. 114–115; "Local Authorities Halt Arctic Nuclear Plant Work," *FBIS*, FBIS-SOV-90-038 (February 26, 1990), p. 77.

73. Ibid.

74. "Balakovo Protests," p. 109.

75. "Protests against Balakovo AES," *FBIS*, FBIS-SOV-90-125 (June 28, 1990), p. 107.

76. "Greens' Protest Halts Construction at Volga AES," *FBIS*, FBIS-SOV-90-228 (November 27, 1990), p. 58.

77. "Russia Okays Plan," p. 12.

78. See Appendix B.

79. "Rally Protests Construction of Khmel'nitskiy AES," *FBIS*, FBIS-SOV-90-119 (June 20, 1990), p. 108.

80. "Pickets Block Khmel'nitskaya Atomic Power Station," *FBIS*, FBIS-SOV-90-140 (July 20, 1990), p. 100.

81. "Datafile: ex-USSR," p. 38.

82. "Public Protests at Nikopol'," pp. 58–59; "Ukrainian Nuclear Power Plant Picketed," *FBIS*, FBIS-SOV-90-113, p. 114.

83. "Construction of Most Nuclear Plants Suspended," *FBIS*, FBIS-SOV-90-237 (December 10, 1990), p. 73.

84. "Datafile: ex-USSR," p. 38.

85. *Izvestiya*, August 2, 1990, p. 2.

86. Ibid.

87. "Construction of Chernobyl' Power Sets Cancelled," *FBIS*, FBIS-SOV-89-076 (April 21, 1989), p. 90.

88. "Russia Okays Plan," pp. 1,12.

89. "Kostroma Atomic Power Plant Construction Halted," *FBIS*, FBIS-SOV-90-130 (July 6, 1990), p. 62.

90. Ibid.

Appendix D

Reactors and Stations in Operation or under Construction in the Newly Independent States of the Former USSR, December 1992

Station	Location	Reactor	Power capacity (MW)	Heating capacity (Gcal/h)	Reactor type	Construction start	Start-up date
Russian Federation							
Beloyarsk AES	Zarechnyy Yekaterinburg Ob.	3	600	280a	BN-600	1966	1980
Novovoronezh AES	Novovoronezhskiy Voronezh Ob.	3	417		VVER-440	1967	1971
		4	417		VVER-440	1967	1972
		5	1,000	100a	VVER-1000	1974	1980
Sosnovy Bor AES	Sosnovy Bor St. Petersburg Ob.	1	1,000/700b		RBMK-1000c	1970	1973
		2	1,000/700b		RBMK-1000c	1970	1975
		3	1,000		RBMK-1000	1973	1979
		4	1,000		RBMK-1000	1975	1981
Kola AES	Polyarnyye Zori Murmansk Ob.	1	440	25	VVER-440c	1970	1973
		2	440	25	VVER-440c	1973	1974
		3	440	25	VVER-440	1977	1981
		4	440	25	VVER-440	1976	1984
Bilibino ATETs	Bilibino Chukotkaya Ob.	1	12	25	EGP-6	1970	1974
		2	12	25	EGP-6	1970	1974
		3	12	25	EGP-6	1970	1975
		4	12	25	EGP-6	1970	1976
Smolensk AES	Desnogorsk Smolensk Ob.	1	1,000		RBMK-1000	1975	1982
		2	1,000		RBMK-1000	1976	1985
		3	1,000	300a	RBMK-1000	1984	1990

Appendix D, Continued

Station	Location	Reactor	Power capacity (MW)	Heating capacity (Gcal/h)	Reactor Type	Construction Start	Start-up Date
Kursk AES	Kurchatov Kursk Ob.	1	1,000/700a		RBMK-1000a	1972	1976
		2	1,000/700a		RBMK-1000a	1973	1979
		3	1,000		RBMK-1000	1978	1983
		4	1,000	350a	RBMK-1000	1981	1985
		5	1,000		RBMK-1000d	1985	UC
Kalinin AES	Udomlya Kalinin Ob.	1	1,000		VVER-1000	1977	1984
		2	1,000	260a	VVER-1000	1982	1986
		3	1,000		VVER-1000	1985	UC
Balakovo AES	Balakovo Satatov Ob.	1	1,000		VVER-1000	1980	1985
		2	1,000		VVER-1000	1981	1987
		3	1,000		VVER-1000	1982	1988
		4	1,000		VVER-1000	1984	UC
South Urals AES	NA Chelyabinsk Ob.	1	800		BN-800	NA	UC
		2	800		BN-800	NA	UC
Voronezh AST	Voronezh Voronezh Ob.	1	heat only	430	AST-500	NA	UC
		2	heat only	430	AST-500	NA	UC
Ukraine Chernobyl' AES	Pripyat Kiev Ob.	1	1,000/700b		RBMK-1000c	1972	1977
		2	1,000/700b		RBMK-1000c	1973	1978
		3	1,000		RBMK-1000	1977	1981
Rovno AES	Kuznetsovsk Rovno Ob.	1	402		VVER-440	1976	1980
		2	416		VVER-440	1977	1981
		3	1,000	100b	VVER-1000	1981	1986
		4	1,000		VVER-1000	1986	UC
South Ukraine AES	Konstatinovka Ivano-Franovsk Ob.	1	1,000		VVER-1000	1977	1982
		2	1,000	260b	VVER-1000	1979	1985
		3	1,000		VVER-1000	1985	1989

Appendix D, Continued

Station	Location	Reactor	Power capacity (MW)	Heating capacity (Gcal/h)	Reactor type	Construction start	Start-up date
Zaporozh'ye AES	Energodar, Zaporozh'ye Ob.	1	1,000	100	VVER-1000	1980	1984
		2	1,000	100	VVER-1000	1981	1985
		3	1,000	100	VVER-1000	1982	1986
		4	1,000	100	VVER-1000	1984	1986
		5	1,000	100e	VVER-1000	1985	1989
		6	1,000	100e	VVER-1000	1986	UC
Khmel'nitsky AES	Neteshin, Khmel'nitskiy Ob.	1	1,000		VVER-1000	1981	1987
		2	1,000		VVER-1000	1985	UC
Lithuania							
Ignalina AES	Snieckus, Lithuania	1	1,500/1250b		RBMK-1500	1977	1983
		2	1,500/1250b		RBMK-1500	1978	1987
Kazakhstan							
Shevchenko ATETs	Shevchenko, Mangyshlak Ob.	1	150		BN-350	1967	1973

a Total capacity reported for station. Smolensk AES total estimated. Beloyarsk AES total probably lower, the figure given includes Beloyarsk-2 reactor that was subsequently shut down.
b Power decreased to lower figure as mandated by *Gospromatomenergo* as of June 1990.
c First-generation reactor of respective type.
d Improved RBMK-1000 incorporating enhanced safety design.
e Estimated by author.

Notes: Location refers to nearest town, and the oblast' (Ob.) in which station is located. Construction start refers to when work began on reactor foundation. Start-up date refers to year in which reactor was connected to the power grid for power supply. UC, under construction.

Sources: IAEA, *Operating Experience*, 1988, pp. 461–549; IAEA, *Operating Experience*, 1992, pp.123–210; N. M. Sinev and B. B. Baturov, *Ekonomika atomnoy energetiki* (Moscow: Energoatomizdat, 1984), p. 51; Petros'yants, *Atomnaya nauka*, pp. 63–64; "Datafile: ex-USSR," p. 37; "Russia Okays Plan," pp. 12–13.

Glossary

AES - Nuclear power station designed specifically for the generation of electricity. From the Russian acronym *atomnaya elektricheskaya stantsiya* (nuclear power station).

AN SSSR - USSR Academy of Sciences. From the Russian acronym for *Akademiya Nauk Soyuza Sovetskikh Sotsialisticheskikh Respublik*.

ASSR - Autonomous Soviet Socialist Republic. The equivalent Russian acronym is ASSR for *Avtonomnaya Sovetskaya Sotsialisticheskaya Respublika*.

AST - Nuclear power station designed specifically for the generation of hot water and steam for domestic, industrial, and municipal use. From the Russian acronym for *atomnaya stantsiya teplosnabzheniya* (nuclear heat supply station).

ATETs - Nuclear power station designed specifically for the cogeneration of electricity and hot water or steam. From the Russian acronym *atomnaya teplovaya elektricheskaya tsentral'* (nuclear heat and power station).

Atomenergostroy - Administration for the Construction of Atomic Power Stations.

CIS - Commonwealth of Independent States. A loose confederation, formed in December of 1991, comprising of most of the newly independent states of the former Soviet Union.

CPSU - Communist Party of the Soviet Union.

Derzhatomnaglyad - Ukrainian State Committee for Nuclear and Radiation Safety. Formed by independent Ukrainian Republic in 1992 to replace *Gosatomnadzor*.

ENIN - Krzhizhanovskiy Energy Institute. From the Russian acronym for *Energeticheskiy institute*. The full title of which is *Gosudarstvennyy nauchno-issledovatel'skiy energeticheshskiy* *institut imeni G. M. Krzhizhanovskogo* (G. M. Krzhizhanov-skiy State Scientific Research Energy Institute).

FBR - Fast-breeder reactor.

Gidroproyekt - All-Union Hydroproject Planning, Surveying, and Research Institute.

GKAE - USSR State Committee for the Utilization of Atomic Energy. From the Russian acronym for *Obshchesoyuznyy gosudarstvennyy komitet SSSR po ispol'zovaniyu atomnoy energii.*

GKNT - USSR State Committee for Science and Technology. From the Russian acronym for *Obshchesoyuznyy gosudarstvennyy komitet SSSR po nauke i tekhnike.*

Glavatomenergostroy - Main Administation for Atomic Power Station Construction.

GLWR - Graphite light-water reactor.

GOELRO - State Commission on the Electrification of Russia. From the Russian acronym for *Gosudarstvennaya komissiya po elektrifikatsii Rossii.*

Gosatomenergonadzor - USSR State Nuclear Power Inspectorate. Created in 1983 replacing the role of *Gosatomnadzor, Gossannnadzor,* and *Gosgortekhnadzor* in the safety regulation of the Soviet nuclear power industry.

Gosatomnadzor - USSR State Nuclear Inspectorate. Created in 1973 to monitor equipment and operational safety in the nuclear power industry.

Gosatomnadzor **(RF)** - State Nuclear Inspectorate of the Russian Federation. Created by the Russia Federation in 1992 to oversee safety in the nuclear industry.

Gosgortekhnadzor - USSR State Industrial and Mining Technical Inspectorate. Until 1983, responsible for operator and reactor management safety

Gosplan - USSR State Committee for Planning .

Gospromatomnadzor - USSR State Nuclear Industry Inspectorate. Created in 1989 formally replacing *Gosatomenergonadzor*.

Gossannadzor - USSR State Sanitation Inspectorate of the Ministry of Health. Until 1983 responsible for on-site monitoring of radiation, environmental impacts and siting in the nuclear industry.

Gosstandart - USSR State Committee of Standards.

Gosstroy -- USSR State Committee for Construction.

Gostekknadzor - USSR State Technical Inspectorate.

GRES - State district thermal power station. From the Russian acronym for *Gosudarstvennaya rayonnaya elektrichesheskaya stantsiya*.

IAEA - International Atomic Energy Agency.

KATEP - Kazakh State Atomic Power Engineering and Industry Corporation

KWU - Kraftwork Union.

LNG - Liquified natural gas.

Minatom - Russian Ministry of Atomic Energy. Created by the Russian Federation in July 1992, replacing the previous Soviet All-Union Ministry *Minatomenergoprom*.

Minatomenergo - USSR Ministry of Nuclear Power. Created in July 1986.

Minatomenergoprom - USSR Ministry of Nuclear Power and Nuclear Industry. Created in July 1989 replacing *Minatomenergo*.

Minenergo - USSR Ministry of Power and Electrification.

Minenergomash - USSR Ministry of Power Machine Building.

Minsredmash - USSR Ministry of Medium Machine Building.

MPA - Maximum design accident. Term used by Soviet reactor designers to describe the most serious potential accident taken into account by reactor designers. From the Russian acronym for *maksimal'naya proektnaya avariya.*

NIMBY - "Not In My Back Yard" Term commonly used to describe negative public attitudes towards noxious facilities that produce a public good or service yet generate negative externalities for the surrounding community or communities.

Oblast' - Region or Province. Administrative area comparable in scale to a U.S. state.

Politburo - Institution representing the highest leadership of the Communist Party of the Soviet Union.

PRA - Probalistic risk assessment.

Promenergoproyekt - All-Union Industrial Energy Planning Institute.

Rayon - District. An administrative unit within an Oblast', republic or large city.

RBMK - Nuclear reactor that uses water as a coolant and graphite as a moderator. From the Russian acronym *reaktor bol'shoi moshchnosti kanal'nyye* (large-capacity channel reactor).

Referentura - An advisory committee to the USSR Council of Ministers that provided research on various issues.

Rosenergoatom - A consortium of Russian operating powerplants under *Minatom.* Created in 1992.

RSFSR - Russian Soviet Federated Socialist Republic. In Russian *RSFSR* for *Rossiyskaya Sovetskaya Federativnaya Sotsial-isticheskaya Respublika.*

Rybnadzor - USSR State Inspectorate of Fisheries.

Sanepidemnadzor **(RF)** - Health and Epidemiological Inspectorate of the Russian Federation.

SO AN SSSR - Siberian Branch of the USSR Academy of Sciences. From the Russian acronym for *Sibirskiy Otdeleniye Akademii Nauk Soyuza Sovetskikh Sotsialistiicheskikh Respublik.*

SSR - Soviet Socialist Republic. Term denoting republican status (e.g. Ukrainian SSR). From the Russian acronym for *Sovetskaya Sotsialisticheskaya Respublika.*

SSSR - USSR or Union of Soviet Socialist Republics. From the Russian acronym for *Soyuz Sovetskikh Sotsialisticheskikh Respublik.*

Teploelektroproyekt - All-Union State Thermal Power Station Planning Institute.

TETs - Thermal cogeneration station. From the Russian acronym *Teplovaya elektricheskaya tsentral'.*

Ukratomenergoprom - A consortium of Ukrainian operating power plants.

VATESI - Nuclear Safety Inspectorate. Created by Lithuania after achieving independence in 1992.

VNIIE - All-Union Scientific Research Institute of Electric Power Engineering. From the Russian acronym for *Vsesoyuznyy nauchno-issledovatel'skiy institut elektroenergetiki.*

VTGR - High-temperature, helium-cooled reactor. From the Russian acronym for *Vysokotemperaturnyy gelievyy reaktor.*

VTI - Dzerzhinskiy All-Union Institute of Heat Engineering. From the Russian acronym for *Vsesoyuznyy teplotekhnicheskiy institut imeni F. E. Dzerzhinskogo.*

VVER - Nuclear reactor that uses water as both a coolant and moderator. Analogous to the Western pressurized, water reactor. From the Russian acronym *vodo-vodyanoy energeticheskiy reaktor* (water-cooled, water-moderated reactor).

WANO - World Association of Nuclear Operators.

Bibliography

Newspapers, Media Articles, and Trade Publications

"Armenia-2, Old VVER-440 is to be Readied for 1992 Restart." *Nucleonics Week*, Vol. 33, No. 2 (January 9, 1992), p. 4.

"Armenian Nuclear Power Station Safety Examined," *FBIS*, FBIS-SOV-88-250 (December 29, 1988), p. 25.

"Atomic Power: Is the Impetus Being Lost?" *CDSP*, Vol. 40, No. 36 (October 5, 1988), pp. 8–9.

"Atomic Power Minister Cited on Safety Measures." *FBIS*, FBIS-SOV-88-027 (February 10, 1988), p. 71.

"Atomic Power Plants Under Way: No Hazards Seen." *CDSP*, Vol. 23, No. 5 (March 2, 1971), p. 25 (from *Ogonyok*, No. 51, December 1970, pp. 6–7).

"Atomic Reactor: Light and Heat." *CDSP*, Vol. 27, No. 28 (August 11, 1976), pp. 18–19 (from *Pravda*, July 14, 1976, p. 6).

"Atomnaya energetika—Nadezhdy vedomstv i trevodi obshchstva." *Novy Mir*, No. 4 (March 1989), pp. 184–193.

"Balakovo Protests against Nuclear Power Station." *FBIS*, FBIS-SOV-90-113 (June 12, 1990), p. 109.

"Bashkiriya Supreme Soviet Debates Nuclear Plant." *FBIS*, FBIS-SOV-90-172 (September 5, 1990), p. 117.

"Belorussian Public Rejects Planned Nuclear Plant." *FBIS*, FBIS-SOV-88-239 (December 13, 1988), p. 63.

Board, William. "Experts Say Reactor Design is 'Immune' to Disaster." *New York Times* (November 15, 1988), pp. 25–26.

187

"Chelyabinsk Nuclear Power Station Planned." *FBIS*, FBIS-SOV-90-238 (December 11, 1990), p. 69.

"Chernobyl' Aftermath: Trees Recovering but Fallout of Faith Remains." *Nucleonics Week*, Vol. 30, No. 12 (March 23, 1989), pp. 6–9.

"Claim of Reactor Subsidence Denied." *FBIS*, FBIS-SOV-88-114 (June 14, 1988), p. 34.

"Conference in the CPSU Central Committee." *CDSP*, Vol. 33, No. 33 (September 16, 1981), p. 19 (from *Pravda*, July 16, 1981, p. 2).

"Construction of Bashkiriya AES Halted." *FBIS*, FBIS-SOV-90-171 (September 4, 1990), p. 106.

"Construction of Bashkiriya AES Halted." *FBIS*, FBIS-SOV-90-237 (December 10, 1990), p. 81.

"Construction of Chernobyl' Power Sets Cancelled." *FBIS*, FBIS-SOV-89-076 (April 21, 1989), p. 90.

"Cuomo Wants NRC to Terminate Shoreham Full-Power License Proceeding." *Nucleonics Week*, Vol. 27, No. 22 (May 22, 1986), p. 7.

"Datafile: ex-USSR." *Nuclear Engineering International*, August 1992, pp. 36–40.

"Deputy Minister on Public Concern Over AES Contamination of Waterways." *JPRS*, JPRS-UPA-89-045 (July 20, 1989), pp. 44–46.

"Differing Views on Ukrainian Stations." *FBIS*, FBIS-SOV-88-032 (February 18, 1988), p. 88.

"The Directive for the Five-Year Plan—Resolution of the 24th Party Congress of the CPSU Central Committee's Draft Directives of the 24th CPSU Congress for the Five Year Plan for the Development of the National Economy in 1971–1975." *CDSP*, Vol. 23, No. 18 (June 1, 1971), p. 13 (from *Pravda*, April 11, 1971, p. 1).

"European Finance Heads Want More Eastern Reactor Safety Upgrades." *Nucleonics Week*, Vol. 33, No. 17 (April 23, 1992), pp. 1, 9.

"Estonia Decides Not to Build Nuclear Plant." *FBIS*, FBIS-SOV-90-224 (November 20, 1990), p. 58.

"Experts Say Human Error May Have Led to Leningrad-3 Failure." *Nucleonics Week*, Vol. 33, No. 31 (July 30, 1992), p. 11.

"Fewer Plants but More Capacity in 1990." *Nuclear Engineering International* (April 1991), p. 4.

"French and German Regulators to Aid Ukrainian Nuclear Safety." *Nucleonics Week*, Vol. 33, No. 35 (August 27, 1992), pp. 9–10.

"From the Stenograph of the Congress of People's Deputies." *FBIS*, FBIS-SOV-89-142-S (July 26, 1989), p. 21.

"Germans and Soviets Agree to Cooperate on Modular HTRs." *Nucleonics Week*, Vol. 29, No. 43 (October 27, 1988), pp. 1, 8–9.

"Gor'kiy Soviet to End Power Station Project." *FBIS*, FBIS-SOV-90-104 (May 12, 1990), p. 83.

"Government Postpones Rostov AES Commissioning." *FBIS,* FBIS-SOV-90-039 (February 27, 1990), p. 93.

"Greens' Protest Halts Construction at Volga AES." *FBIS*, FBIS-SOV-90-228 (November 27, 1990), p. 58.

"Group Calls for Blockade of Plant." *FBIS*, FBIS-SOV-88-176 (September 14, 1988), p. 56.

"Guidelines for the 10th Five-Year Plan—II." *CDSP*, Vol. 28, No. 16 (May 16, 1976), pp. 9–12 (from *Izvestiya*, March 7, 1976, pp. 2–8).

"Guidelines for the 11th Five-Year Plan." *CDSP*, Vol. 33, No. 16 (May 20, 1981), p. 11 (from *Izvestiya*, March 5, 1981, p. 3).

"How Likely is Another Chernobyl'?" *CDSP*, Vol. 40, No. 42 (November 16, 1988), pp. 1–6.

"IAEA Conference Repeats Promises but Produces No Safety Aid to East," *Nucleonics Week*, Vol. 33, No. 43 (October 22, 1992), pp. 7–8.

"In Armenia, Elsewhere." *FBIS*, FBIS-SOV-88-248 (December 27, 1988), p. 67.

"Industrial and Nuclear Power Safety Chairman Approved on the 14th." *BBC, SWB*, SU/0511 (July 18, 1989), C/8.

Izvestiya. September 7, 1988, p. 2.
  October 28, 1989, p. 3.
  November 27, 1989, p. 2.
  August 2, 1990, p. 2.
  August 11, 1990, p.2.
  August 25, 1990, p. 2.
  October 16, 1990, p. 6.
  January 28, 1991, p. 3.

"Jury Gives Russia Choice of 11 Viable Reactor Designs." *Nucleonics Week*, Vol. 33, No. 23 (June 4, 1992), p. 13.

"Karelian Nuclear Project Suspended." *FBIS*, FBIS-SOV-90-042 (March 2, 1990), p. 84.

"Kazakhstan Plans New FBR at Shevchenko Research Complex." *Nucleonics Week*, Vol. 33, No. 44 (October 29, 1992), pp. 7–8.

Komsomol'skaya pravda. May 24, 1990, p. 2.

"Kostroma Atomic Power Plant Construction Halted." *FBIS*, FBIS-SOV-90-130 (July 6, 1990), p. 62.

"Krasnodar Public Opinion That Caused Cancelling of Nuclear Power Station Laid to Uninformed Post-Chernobyl' Fear." *CDSP*, Vol. 40, No. 3 (February 17, 1988), p. 9 (from *Komsomolskaya pravda*, January 27, 1988).

"Leaders Address 27th Party Congress—VII." *CDSP*, Vol. 38, No. 15 (May 14, 1986), p. 10 (from *Pravda*, March 5, 1986, p. 3).

Literaturna Ukraina. June 23, 1988, p. 4.

Literaturnaya gazeta. November 2, 1988, p. 1.

"Lithuanians Face Myriad Problems at Ignalina Nuclear Station." *Nucleonics Week*, Vol. 33, No. 16 (April 16, 1992), pp. 7–8.

"Lithuanians See Possible Swedish Models for Fuel Storage, Outages." *Nucleonics Week*, Vol. 33, No. 36 (September 3, 1992), p. 12.

"Local Authorities Halt Arctic Nuclear Plant Work." *FBIS*, FBIS-SOV-90-038 (February 26, 1990), p. 77.

"Local Authorities Will Make Decisions on Building Nuclear Power Plants." *CDPSP*, Vol. 44, No. 26 (July 29, 1992), p. 34.

Los Angeles Times. August 26, 1986, p. 6.

"Minister Discusses Nuclear Power Development." *FBIS*, FBIS-SOV-88-248, (December 27, 1988), p. 67.

"Minister Says Talk of Closing Older RBMKs is 'Pure Fantasy'." *Nucleonics Week*, Vol. 33, No. 39 (September 24, 1992), p. 14.

"Moscow Regional Center Reading as USSR Prepares for WANO Opening." *Nucleonics Week*, Vol. 30, No. 14 (March 9, 1989), pp. 3–4.

"New Nuclear Ministry Has Been Formed in the USSR." *Nuclear News*, August 1989, p. 19.

"No Feasible Alternative to Atomic Power." *CDSP*, Vol. 39, No. 17 (May 27, 1987), pp. 5, 9.

"Nuclear News Briefs." *Nuclear News*, August 1986, p. 29.

"Nuclear Power Station Causes Controversy." *FBIS*, FBIS-SOV-89-194, (October 10, 1989), p. 91.

"Officials Challenged on Ecological Safety in Projects near Kazan and Astrakhan." *BBC, SWB*, SU/0369 (January 27, 1989), C2/1.

"Ohio Governor Pulls Support for Perry and Davis-Besse Emergency Plans." *Nucleonics Week*, Vol. 27, No. 34 (August 21, 1986), pp. 1–2.

"100,000 Sign Nuclear Power Plant Protest Message." *FBIS*, FBIS-SOV-89-146 (August 1, 1989), p. 81.

"Opposition to Nuclear Power Stations in the Ukraine." *BBC, SWB*, SU/0451 (May 5, 1989), C/3.

"Pickets Block Khmel'nitskiy Atomic Power Station." *FBIS*, FBIS SOV-90-140 (July 20, 1990), p. 100.

Pravda. March 1, 1986, p. 2.
  January 21, 1988, p. 6.
  January 11, 1989, p. 3.
  February 1, 1989, pp. 2–3.
  October 5, 1989, p. 6.
  January 19, 1990, p. 2.
  July, 19, 1990, p. 2.

Pravda ukrainy. January 15, 1989, pp. 1, 3.

"Precautions in Expanding Nuclear Power Viewed." *FBIS*, FBIS-SOV-88-046 (March 9, 1988), p. 47.

"Protests against Balakovo AES." *FBIS*, FBIS-SOV-90-125 (June 28, 1990), p. 107.

"Protest against Karelian Power Plant Construction." *FBIS*, FBIS SOV-90-012 (January 18, 1990), p. 132.

"Public Protests at Nikopol'." *FBIS*, FBIS-SOV-90-132 (July 10, 1990), pp. 58–59.

"Rally Protests Tatar AES Construction." *FBIS*, FBIS-SOV-90-123 (June 26, 1990), p. 105.

"Russia Blocking Consensus to Shut RBMKs, Germans Say." *Nucleonics Week,* Vol. 33, No. 31 (July 30, 1992), pp. 8–9.

"Russia Okays Plan to Proceed with Major Nuclear Construction." *Nucleonics Week,* Vol. 34, No. 3 (January 21, 1993) pp. 1, 12–13.

"Russian Government is Very Short of Energy." *CDSP,* Vol. 44, No. 22 (July 1, 1992), p. 22.

"Russians Favor VVER-1000 as Replacement Generation at Kola." *Nucleonics Week,* Vol. 33, No. 26 (June 25, 1992), p. 13.

"Russians Losing Hope that Nuclear Aid Will Materialize from West." *Nucleonics Week,* Vol. 33, No. 30 (July 23, 1992), pp. 10–13.

"Russians Open Plant Selection to International Jury." *Nucleonics Week,* Vol. 33, No. 12 (March 12, 1992), pp. 1, 7.

"Safety Record Reviewed on Large VVER Units." *Nuclear News,* January 1989, p. 103.

"Seabrook Evacuation Plans Become Election Issue in Two States." *Nucleonics Week,* Vol. 31, No. 20 (May 15, 1990), p. 9.

"Seabrook Startup Delays Have Cost $300 Million." *Nucleonics Week,* Vol. 27, No. 23 (May 29, 1986), p. 14.

"Seimens/GRS Get First Russian VVER Risk Assessment Contract." *Nucleonics Week,* Vol. 29, No. 39 (September 29, 1988), pp. 2–3.

"Sidorenko Says Minatom-Financed Backfits of RBMKs Will Continue." *Nucleonics Week,* Vol. 33, No. 19 (May 7, 1992), p. 4.

Sotsialisticheskaya industriya. March 26, 1989, p. 3.
June 10, 1989, p. 4.
October 21, 1989, p. 1

Sovetskaya rossiya. April 26, 1989, p.3.
October 22, 1989, p. 1.
April 1, 1990, p. 1.
May 21, 1990, p. 3.

"Soviet Nuclear Power Plans Outlined." *CDSP,* Vol. 33, No. 22 (June 21, 1981) pp. 5–6, (from *Pravda,* June 4, 1981, p. 2).

"Soviets Developing Enhanced RBMK." *Nuclear Engineering International,* July 1988, p. 2.

"Soviets Report Development of 'Enhanced', Safer RBMK-1500." *Nucleonics Week,* Vol. 29, No. 12 (March 24, 1988), pp. 1, 10.

"Tatar Anti-Nuclear Protests Reported." *FBIS,* FBIS-SOV-89-146 (August 1, 1989), pp. 79–80.

"U.K./U.S.S.R.: Regulators Agree to Information Exchange." *Nucleonics Week,* Vol. 29, No. 14 (April 7, 1988), p. 11.

"Ukrainian Lawmakers Striving to Limit Nuclear Utility Power." *Nucleonics Week,* Vol. 33, No. 40 (October 8, 1992), pp. 12–13.

"Ukrainian Nuclear Power Plant Picketed." *FBIS,* FBIS-SOV-90-113 (June 12, 1990), p. 114.

"Understanding the CIS: Lord Marshall's View," *Nuclear Engineering International,* September 1992, pp. 15–25.

"USSR Seeks Public Acceptance of Improved Reactor Designs." *Nucleonics Week,* Vol. 29, No. 39 (September 29, 1988), pp. 1, 8–9.

"USSR to Close Armenian Nuclear Plant to Meet Public Demands." Nucleonics Week, Vol. 29, No. 52 (December 15, 1988), p. 1.

"Work Halted on Arkhangel'sk Nuclear Power Plant." *FBIS,* FBIS-SOV-90-035 (February 21, 1990), pp. 114–115.

"Workers Protest Lost Jobs at Rostov Nuclear Plant." *FBIS,* FBIS-SOV-90-220 (November 14, 1990), p. 45.

Books, Journals, and Statistical Handbooks

Abagyan, A. A., et al. "Nuclear Power in the USSR." *Soviet Atomic Energy,* Vol. 69, No. 2 (February, 1991), pp. 621–629 (from *Atomnaya Energiya,* Vol. 69, No. 2, August 1990, pp. 67–79).

Abramov, V. M. "The Bilibino Nuclear Power Station." *Soviet Atomic Energy,* Vol. 35, No. 5 (May 1974), pp. 977–980 (from *Atomnaya Energiya,* Vol. 35, No. 5, November 1973, pp. 299–304).

Aleksandrov, A. P. "Nuclear Power Problems." *Soviet Atomic Energy,* Vol. 13, No. 2 (March, 1963), pp. 710–718 (from *Atomnaya Energiya,* Vol. 13, No. 2, August 1962, pp. 109–124).

Aleksandrov, A. P. *Yadernaya energetika, chelovek i okruzhayushchaya sreda.* Moscow: Energoatomizdat, 1984.

Aleksashin, P. P., et al. "Razvitiye trebovaniy po bezopasnosti i sistemi gosudarstvennnogo nadzora kak osnovi bezopasnogo razvitiya yadernoy energetiki." In *Nuclear Power Performance and Safety,* Vol. 4, pp. 427–434. Vienna: IAEA, 1987.

Anan'yev, E. P. and G. N. Kruzhilin. "Radioactive Safety Barriers in Nuclear Power Stations." *Soviet Atomic Energy,* Vol. 37, No. 1 (July, 1974), pp. 679–705, (from *Atomnaya Energiya,* Vol. 37, No. 1, January 1974, pp. 369–373).

Asmolov, V. G., et al. "The Chernobyl' Accident: One Year Later." *Soviet Atomic Energy,* Vol. 64, No. 1 (July, 1988), pp. 4–26, (from *Atomnaya Energiya,* Vol. 64, No. 1, January 1988, pp. 2–24).

Asmolov, V. G., et al. "Rol' reglamentipuyushchykh polozheniy v povyshenii urovnya bezopasnosti atomnykh stantsiy." In *Regulatory Practices and Safety Standards for Nuclear Power Plants,* pp. 547–464. Vienna: IAEA, 1989.

Bahry, Donna. *Outside Moscow: Power, Politics, and Budgetary Policy in the Soviet Republics.* New York, NY: Columbia University Press, 1987.

Belyaev, A. I. "USSR Regulatory and Supervisory Practices in Nuclear Plant Safety." In *Regulatory Practices and Safety Standards for Nuclear Power Plants,* pp. 53–58. Vienna: IAEA, 1989.

Berkovich, V. M., et al. "Designing a Nuclear Power Station with 1000 Mw Water-Moderated Water-Cooled Reactor Units." *Thermal Power Engineering,* Vol. 21, No. 4 (April 1974), pp. 26–29 (from *Teploenergetika,* Vol. 21, No. 4, April 1974, pp. 18–22).

Bukrinskii, A. M., et al. "Nuclear Plant Safety and Govenment Regulation." Soviet Atomic Energy, Vol. 68, No. 5 (November 1990), pp. 380–385 (from *Atomnaya Energiya,* Vol. 68, No. 5, May 1990, pp. 328–332).

Campbell, Robert W. *Soviet Energy Technologies: Planning, Policy, Research and Development.* Bloomington, IN: Indiana University Press, 1980.

Chung, Han-Ku. *Interest Representation in Soviet Policymaking: A Case Study of a West Siberian Energy Coalition.* Boulder, CO: Westview Press, 1987.

CIA. *CIA Handbook of Economic Statistics.* Washington, D.C.: Central Intelligence Agency, 1975.

CIA. *CIA Handbook of Economic Statistics.* Washington, D.C.: Central Intelligence Agency, 1990.

CIA. *USSR Energy Atlas.* Washington, D.C.: Central Intelligence Agency, 1985.

Cook, Constance Ewing. *Nuclear Power and Legal Advocacy.* Lexington, MA: Lexington Books, 1980.

Darst, Robert G. "Environmentalism in the USSR: The Opposition to the River Diversion Projects." *Soviet Economy,* Vol. 4, No. 3 (July–September 1988), pp. 223–252.

Del Sesto, Steven L. *Science, Politics and Controversy: Civilian Nuclear Power in the United States, 1946–1974.* Boulder, CO: Westview Press, 1979.

Dergachev, N. P. "Kriterii vybora ploshchadok dlya issledovatel'skikh tsentov, energeticheskikh ustanovok i ustanovok toplivnogo tsikla v SSSR." In *Siting of Nuclear Facilities,* pp. 79–84. Vienna: IAEA, 1975.

Dienes, Leslie. "The Energy System and Economic Imbalances in the USSR." *Soviet Economy,* Vol. 1, No 4 (October–December 1985), pp. 340–372.

Dienes, Leslie, and Theodore Shabad. *The Soviet Energy System: Resource Use and Policies.* Washington, D.C.: V. H. Winston and Sons, 1979.

Dierkes, M., et al. (eds.). *Technological Risk: Its Perception and Handling in the European Community.* Boston, MA: Oetgeschlager, Gunn, and Hain, 1979.

Dodd, Charles K. *Siting Hazardous Facilities in the Soviet Union: The Case of the Nuclear Power Industry.* Master's thesis, University of Washington, 1992.

Dollezhal', N. A., and Yu. I. Koryakin. "Nuclear Energy Complexes and the Economic and Ecological Problems of Nuclear Power Development." *Soviet Atomic Energy,* Vol. 43, No. 5 (November 1977), pp. 1019–1024 (from *Atomnaya Energiya,* Vol. 43, No.5, May 1977, pp. 369–373).

Dollezhal', N. A., and Yu. I. Koryakin. "Yadernaya elektroenergetika: dostizheniye i problemy." *Kommunist,* No. 14 (September 1979), pp. 19–28.

Dollezhal', N. A., and I. Ya. Yemel'yanov. "Experience in the Construction of Large Power Reactors in the USSR." *Soviet Atomic Energy,* Vol. 40, No. 2 (August 1976), pp. 143–149 (from *Atomnaya Energiya,* Vol. 40, No. 2, February 1976, pp. 117–126).

Donahue, M., et al. "Assessment of the Chernobyl'-4 Accident Localization System." *Nuclear Safety,* Vol. 28, No. 3 (July–September 1987), pp. 297–310.

Dooley, James E., et al. "The Management of Nuclear Risk in Five Countries: Political Cultures and Institutional Settings." In Roger E. Kasperson and Jeanne X. Kasperson (eds.). *The Impacts of Large-Scale Risk Assessment in Five Countries,* pp. 27–48. Boston, MA: Allen & Unwin, 1987.

Dyker, David. *The Process of Investment in the Soviet Union*. Cambridge: Cambridge University Press, 1983.

Ebbin, Steven, and Raphael Kasper. *Citizen Groups and the Nuclear Power Controversy: Uses of Scietific and Technological Information*. Cambridge, MA: MIT Press, 1974.

Fedorenko, N. P. (ed.). *Programno-tselevoy metod v planirovanii*. Moscow: Nauka, 1982.

Fernie, John, and Stan Openshaw. "Policy Making and Safety Issues in the Development of Nuclear Power in the United Kingdom." In Pasqualetti and Pijawka, pp. 67–92.

Flakserman, Yu. N. *Teploenergetika SSSR, 1921–1980*. Moscow: Nauka, 1981.

Freudenburg, William, and Eugene A. Rosa (eds.). *Public Reactions to Nuclear Power: Are There Critical Masses?* Boulder, CO: Westview Press, 1984.

Fuller, Elizabeth. "Is Nuclear Power the Answer to Georgia's Energy Problems?" *RFE/RL, Radio Libery Research Reports*, RL 282/81 (July 17, 1981), pp. 1–3.

Girnius, Saulius "Continued Controversy over the Third Reactor at Ignalina Atomic Power Plant." *RFE/RL, Radio Liberty–Baltic Area*, SR 18 (August 4, 1988), pp. 29–32.

Girnius, Saulius. "The Ignalina Atomic Plant's Second Reactor in Operation." *RFE/RL. Radio Liberty–Baltic Area*, SR/4 (May 4, 1988), pp. 12–13.

Gold, Donna. *Agenda Setting in Soviet Domestic Politics: The Case of Nuclear Safety Policy*. Paper presented to the Annual Conference of the American Association of Slavic Studies, Honolulu, November 18–21, 1988.

Gorin, V. I., and Ya. F. Masalov. "Povysheniye effektivnosti teplovikh elektrostanstiy." *Teploenergetika*, Vol. 28, No. 6 (June 1981), pp. 5–8.

Gorshkov, A. S. "Soviet Heat and Power Engineering on the Eve of the 23rd CPSU Congress." *Thermal Power Engineering*, Vol. 13, No. 3 (March 1966), pp. 3–6, (from *Teploenergetika*, Vol. 13, No. 3, March 1966, pp. 2–4).

Goskomstat SSSR. *Kapital'noye stroitel'stvo: statisticheskiy sbornikh*. Moscow: Finansy i Statistika, 1988.

Goskomstat SSSR. *Material'no-tekhnicheskoye obespecheniye narodnogo khozyaystva SSSR*. Moscow: Finansy i Statistika, 1988.

Goskomstat SSSR. *Narodnoye khozyaystvo SSSR v 1989*. Moscow: Finansi i Statistika, 1990.

Goskomstat SSSR. *Narodnoye khozyaystvo SSSR v 1990*. Moscow: Finansi i Statistika, 1991.

Goskomstat SSSR. *Naseleniye SSSR, 1987*. Moscow: Finansy i Statistika, 1988.

Goskomstat SSSR. *Promyshlennost' SSSR*. Moscow: Finansy i Statistika, 1988.

Green, A. E. *High-Risk Safety Technology*. New York, NY: John Wiley & Sons, 1982.

Grimes, B. K. "External Hazards as They Affect Nuclear Power Plant Siting." In *Siting Nuclear Facilities*, pp. 140–152. Vienna: IAEA, 1975.

Gustafson, Thane. *Reform in Soviet Politics: Lessons of Recent Policies on Land and Water*. Cambridge: Cambridge University Press, 1981.

Hewett, Ed A. *Energy Economics and Foreign Policy in the Soviet Union*. Washington, D.C.: The Brookings Institution, 1984.

Hewett, Ed A. *Reforming the Soviet Economy: Equality versus Efficiency*. Washington, D.C.: The Brookings Institution, 1988.

Huzinec, George A. "A Reexamination of Soviet Industrial Location Theory." *The Professional Geographer*, Vol. 29, No. 3 (August 1977), pp. 259–265.

IAEA. *Operating Experience with Nuclear Power Stations in Member States in 1987*. Vienna: IAEA, 1988.

IAEA. *Operating Experience with Nuclear Power Stations in Member States in 1990*. Vienna: IAEA, 1991.

Ignatenko, Ye. I. "Osnovniye napravleniya rabot po povyshenko bezopasnosti seruynykh atomnykh energoblokov." *Elektricheskiye stanstii*, No. 8 (August 1989), pp. 21–29.

Il'kevich, A. A., et al. "Investigating the Optimum Unit Capacity and the Composition of the Main Items of Plant for Nuclear Heat and Power Stations." *Thermal Power Engineering*, Vol. 21, No. 2 (February 1974), pp. 4–6 (from *Teploenergetika*, Vol. 21, No. 2, February 1974, pp. 3–7).

Ishikawa, M. "An Examination of the Accident Scenario at the Chernobyl' Nuclear Power Station." *Nuclear Safety*, Vol. 28, No. 4 (October–December 1987), pp. 449–454.

Jackson, Marvin. "Information and Incentives in Planning Soviet Investment Projects." *Soviet Studies*, Vol. 23, No. 1 (July 1971), pp. 3–25.

Jasper, James M. *Nuclear Politics: Energy and the State in the United States, Sweden and France.* Princeton, NJ: Princeton University Press, 1990.

John, T. F. "Environmental Pathways of Radioactivity to Man," In Vol. 3, *Nuclear Power Technology: Radiation.* Oxford: Clarendon Press, 1983, pp. 155–216.

Katsman, David. "Balance of Plant in Soviet VVER-1000 Reactors: The Case of Side-Mounted Condensers. In Young, pp. 32–58.

Katsman, David. *Soviet Nuclear Power Plants: Reactor Types, Water and Chemical Control Systems, Turbines.* Falls Church, VA: Delphic Associates, 1986.

Kazachkovskiy, O. D., et al. "Razvitiye i opit ekspluatsii reaktorov na bistrikh neytronakh v SSSR." In *Nuclear Power Experience*, Vol. 5, pp. 18–24. Vienna: IAEA, 1982.

Keeney, Ralph. *Siting Energy Facilities.* New York, NY: Academic Press, 1982.

Kelley, Donald R. "Environmental Policy-Making in the USSR: The Role of Industrial and Environmental Interest Groups," *Soviet Studies*, Vol. 28, No. 4 (October 1976), pp. 570–589.

Kelly, William, et al. "The Economics of Nuclear Power in the Soviet Union." *Soviet Studies*, Vol. 34, No. 1 (January 1982), pp. 43–68.

Kelly, William, et al. *Energy Research and Development in the USSR: Preparations for the Twenty First Century.* Durham, NC: Duke University Press, 1986.

Kitschelt, Herbert. "Political Oppotunity Structures and Political Protest: Anti-Nuclear Protest in Four Democracies." *British Journal of Political Science*, Vol. 16, No. 1 (January 1986), pp. 57–85.

Kolbasov, D. "Ekologicheskaya politika SSSR." *Sovetskoye gosudarstvo i pravo*, No. 3, 1982, pp. 82–84.

Komarovskii, A. N. *Design of Nuclear Plants.* Jerusalem: Israel Program for Scientific Translations, 1968, Appendix to Part I (translated from the Russian source *Stroitel'stvo yadernikh ustanovok*, Moscow: Atomizdat, 1965).

Koryakin, Yu. I. "Mathematical Modeling of the Developing Nuclear Power Generation." *Soviet Atomic Energy*, Vol. 36, No. 6 (December 1974), pp. 586–587 (from *Atomnaya Energiya*, Vol. 36, No. 6, June 1974, pp. 419–420).

Koryanikov, V. P. "Present-Day Conditions and Prospects for the Development of District Heating." *Thermal Power Engineering*, Vol. 19, No. 4 (April 1972), pp. 1–6, (from *Teploenergetika*, Vol. 19, No. 4, April 1974, pp. 2–5).

Kovylyanskii, Ya. A. "Centralized Heat Supply from Nuclear Sources." *Thermal Power Engineering*, Vol. 28, No. 3 (March 1981), pp. 130–138 (from *Teploenergetika*, Vol. 28, No. 3, March 1981, pp. 10–16).

Kruglov, M. G., et al. "Prioritetnyye napravleniya i gosudarstvennyye programmy nauchno-tekhnicheskogo progressa v proizvodstve i ispol'zovanii energeticheskikh resursov." *Teploenergetika*, Vol. 36, No. 1 (January, 1989), pp. 2–7.

Kunreuther, H., et al. *Risk Analysis and Decision Processes: The Siting of Liquified Energy Facilities in Four Countries.* New York, NY: Springer-Verlag, 1983.

Legasov, V. A., and V. M. Novikov. "Bezopasnost' i effektivnost' yadernoy energetiki: kriteri, puti sovershenstvovaniya." In *Nuclear Power Performance and Safety*, Vol. 2, pp. 447–458, Vienna: IAEA, 1988.

Leipunskiy, A. I., et al. "Construction of an Atomic Electric Power Plant Based on the BN-350 Reactor." *Soviet Atomic Energy*, Vol. 23, No. 5 (November 1967), pp. 1163–1167 (from *Atmonoya Energiya*, Vol. 23, No. 5, May, 1967, pp. 409–416).

Levental', A. A., et al. "Technical Problems of District Heating on the Basis of Nuclear Fuel." *Thermal Engineering*, Vol. 21, No. 11 (November 1974), pp. 16–20 (from *Teploenergetika*, Vol. 21, No. 11, November 1974, pp. 10–16).

Lewin, Joseph. "The Russian Approach to Nuclear Reactor Safety." *Nuclear Safety*, Vol. 18, No. 4 (July–August 1977), pp. 438–450.

Marples, David. *Chernobyl' and Nuclear Power in the USSR.* New York, NY: St. Martin's Press, 1986.

Marples, David. "Chigirin and the Soviet Nuclear Energy Program." *RFE / RL, Report on the USSR*, RL 366/89 (July 9, 1989), pp. 26–29.

Marples, David. "Decree on Ecology Adopted in Ukraine." *RFE / RL, Report on the USSR*, RL 161/90 (March 15, 1990), pp. 15–16.

Marples, David. "New Protests Against Soviet Nuclear Energy Program." *RFE / RL, Radio Liberty Research Reports*, RL 448/88 (September 28, 1988), pp. 4–5.

Marples, David. "Ukraine Declares Moratorium on New Nuclear Reactors." *RFE / RL. Report on the USSR*, RL 425/90 (September 10, 1990), pp. 20–21.

McCormick, William. *Reliability and Risk Analysis: Methods and Nuclear Power Applications.* New York, NY: Academic Press, 1981.

Melent'yev, L. A., and A. A. Makarov. *Energeticheskiy kompleks SSSR.* Moscow: Economika, 1983.

Minenergo. *Razvitiye elektroenergeticheskogo khozyaystva SSSR: Khronologicheskiy ukazatel'.* Moscow: Energoatomizdat, 1987.

Minenergo. *Razvitiye elektroenergetiki soyuznikh respublik.* Moscow: Energoatomizdat, 1988.

Minenergo. "Trebovaniya k razmeshcheniyu atomnykh stanstiy teploshabzheniya i atomnykh teploelektrotsentraley po usloviyam radiatsionnoy bezopasnosti." *Atomnaya Energiya,* Vol. 49, No. 2 (August 1980), pp. 150–151.

Moe, Alrid, and Helge Ole Bergson. "The Soviet Gas Sector: Challenges Ahead." In *The Soviet Economy: A New Course,* pp. 153–194, Brussels: NATO, 1987.

Mounfield, Peter. *World Nuclear Power.* New York, NY: Routledge, 1991.

Mumphrey, A., et al. *A Decision Model for Locating Controversial Facilities.* Discussion Paper No. 11. Philadelphia: University of Pennsylvania, Department of Geography, 1971.

Nahaylo, Bohdan. "More Ukrainian Scientists Voice Opposition to Expansion of Nuclear Energy Program." *RFE / RL. Radio Liberty Research Reports,* RL 135/88 (March 21, 1988), pp. 1–2.

Nekrasov, A. M. *Problemi razvitiye i razmeshcheniya toplivnikh baz SSSR.* Moscow: Nauka, 1982.

Nekrasov, A. M., and A. A. Troitskiy (eds.). *Energetika v SSSR 1981–1985.* Moscow: Energoatomizdat, 1982.

Nelkin, Dorothy, and Micheal Pollak. *The Atom Besieged: Extraparliamentary Dissent in France and Germany.* Cambridge, MA: MIT Press, 1981.

Nelkin, Dorothy, and Micheal, Pollak. "Consensus and Conflict Resolution: The Politics of Assessing Risk." In Diekes et al. (eds.), pp. 65–75.

Neporozhniy, P. S. (ed.). *Stroitel'stvo teplovykhi atomnykh elektrostantsiy.* Moscow: Stroiizdat, 1985.

Nigmatulin, I. I., and B. I. Nigmatulin. *Yadernyye energeticheskiye ustanovk.* Moscow: Energoatomizdat, 1986.

Nikiporets, Yu. G., et al. "Bezopasnost' atomnykh stantsiy teplosnabzheniya v SSSR." In *Nuclear Power Experience*, Vol. 4, pp. 147–158. (Vienna: IAEA, 1983).

Openshaw, Stan. *Nuclear Power: Siting and Safety*. London: Routledge and Kegan Paul, 1986.

Pallot, Judith, and Denis Shaw. *Planning in the Soviet Union*. London: Croon Helm, 1984.

Pasqualetti, M. J., and K. D. Pijawka (eds.). *Nuclear Power: Assesing and Managing a Hazardous Technology*. Boulder, CO: Westview Press, 1984.

Pavlenko, A. S. and A. M. Nekrasov (eds.). *Energetika v SSSR, 1971–1975*. Moscow: Energiya, 1972.

Peterson, D. J. "Hard Times for the Environment." *RFE/RL, Report on the USSR*, RL 397/91 (November 15, 1991), pp. 15–19.

Petros'yants, A. M. *Atomnaya nauka i teknika*. Moscow: Energoatomizdat, 1987.

Petros'yants, A. M., "A Decade of Nuclear Power Engineering." *Soviet Atomic Energy*, Vol. 16, No. 6 (June 1964), pp. 596–601 (from *Atomnaya Energiya*, Vol. 16, No. 6, June 1964, pp. 479–484).

Petros'yants, A. M. *Sovetskoye atomnoye pravo*. Moscow: Nauka, 1986.

Petros'yants, A. M. et al. "The Leningrad Nuclear Power Station and the Outlook for Channel Type BWRs." *Soviet Atomic Energy*, Vol. 31, No. 4 (April 1972), p. 1088 (from *Atomnaya Energiya*, Vol. 31, No. 10, October 1971, p. 317).

Petros'yants, A. M., et al. "Prospects of the Development of Nuclear Power in the USSR." *Soviet Atomic Energy*, Vol. 31, No. 4 (April 1972) pp. 1067–1074 (from *Atomnaya Energiya*, Vol. 31, No. 4, October 1971, pp. 315–323).

Pijawka, David K. "The Pattern of Public Response to Nuclear Facilities: An Analysis of the Diablo Canyon Nuclear Generating Station.," In Pasqualetti and Pijawka, pp. 213–239.

Pijawka, David K. "Public Sector Effects and Social Impact Assessment of Nuclear Generating Facilities: Information for Community Mitigation Management." In Pasqualetti and Pijawka, pp. 171–188.

Price, Jerome. *The Antinuclear Movement*. Boston, MA: Twayne Publishers, 1990.

Pruzner, S. L. *Ekonomika, oganizatsiya i planirovaniye energeticheskogo proizvodstva*. Moscow: Energoatomizdat, 1984.

Prunzer, S. L., et al. *Organizatsiya, planirovaniye i upravleniye energeticheskim predpriyatem.* Moscow: Vysshaya Shkola, 1981.

Raddatis, K. F., and K. V. Shakhsuvarov. "O poteryakh v narodnom khozyaystve iz za ponizhennogo kachestva uglei dlya teplovikh elektrostanstii." *Elektricheskiye stantsii*, No. 1 (January, 1985), pp. 6–10.

Revelle, Charles, et al. "An Analysis of Private and Public Sector Location Models." *Management Science*, Vol. 16, No. 11 (July 1970), pp. 692–707.

Richetto, Jeffery P. "Locating Nuclear Electric Energy Facilities: Structural Relationships and the Environment." In Pasqualetti and Pijawka, pp. 103–121.

Rosa, Eugene, and William Freudenburg. "Nuclear Power at the Crossroads." In Freudenburg and Rosa (eds.), 22–34.

Ryl'skiy, V. A. *Elektroenergeticheskaya baza ekonomicheskikh rayonov SSSR.* Moscow: Nauka, 1974.

Sagers, Matthew. "News Notes." *Post-Soviet Geography*, Vol. 33, No. 4 (April 1992), pp. 237–267.

Sagers, Matthew. "News Notes." *Soviet Geography*, Vol. 28, No. 12 (December 1987), pp. 778–783.

Sagers, Matthew. "News Notes." *Soviet Geography*, Vol. 30, No. 4 (April 1989), pp. 338–352.

Sagers, Matthew. "News Notes." *Soviet Geography*, Vol. 30, No. 5 (May 1989), pp. 420–428.

Sagers, Matthew. "News Notes." *Soviet Geography*, Vol. 32, No. 4 (April 1991), pp. 253–290.

Sagers, Matthew, and Albina Tretyakova. "Constraints in Gas and Oil Substitution in the USSR: The Oil Refining Industry and Gas Storage." *Soviet Economy*, Vol. 2, No. 1 (January–March 1986), pp. 72–94.

Semple, R. Keith, and Jeffery P. Richetto, "The Location of Electric Energy Facilities: Conflict, Coalition and Power." *Regional Science Perspectives.* Vol. 9, No. 1 (January 1979), pp. 117–138.

Semple, R. Keith, and Jeffery, P. Richetto, "Locational Trends of an Experimental Public Facility: The Case of Nuclear Power Plants." *The Professional Geographer*, Vol. 28, No. 3 (August 1976), pp. 248–253.

Shabad, Theodore. "News Notes." *Soviet Geography*, Vol. 19, No. 4 (April 1978), pp. 274–282.

Shabad, Theodore. "News Notes." *Soviet Geography*, Vol. 27, No. 4 (April 1986), pp. 248–279.

Shabad, Theodore. "News Notes." *Soviet Geography*, Vol. 27, No. 7 (September, 1986), pp. 504-525.

Shadrin, A. P.,*Atomniye elektrostatsii na kraynem severe*. Yakutsk: Yakutskiy filial SO AN SSSR, 1983.

Shasharin, A. G. , et al. "State-of-the-Art and Development Prospects for Nuclear Power Stations Containing Pressurized Water Reactors." *Soviet Atomic Energy*, Vol. 56, No. 6 (December 1984), pp. 361-367 (from *Atomnaya Energiya*, Vol. 56, No. 6, June 1984, pp. 353–359).

Sheehy, Ann, and Sergei Voronitsyn. "Ecological Protest in the USSR, 1986–1988." *RFE/RL, Radio Liberty Research Reports*, RL 191/88 (May 11, 1988), pp. 1–3.

Shrader-Frechette, K. S. *Science Policy, Ethics and Economic Methodology: Some Problems of Technology, Assessment and Environmental-Impact Analysis*. Boston, MA: D. Reidel Publishing Co., 1985.

Sidorenko, V. A., et al. "Normirovaniye bezopasnosti atomnikh stantsiy v SSSR." In *Nuclear Power Experience*, Vol. 4, pp. 625–634. Vienna: IAEA, 1982.

Sidorenko, V. A., et al. "Opit sozdaniya, ekspluatatsii i puti sovershenstvovaniya AES c VVER." In *Nuclear Power Experience*, Vol. 2, pp. 51–68. Vienna: IAEA, 1982.

Sidorenko, V. A., et al. "Present Day Problems of Safe Operation of Nuclear Power Stations," Thermal Power Engineering, Vol. 33 No. 3 (March 1976), p. 9 (from Teploenergetika, Vol. 33, No. 3, March 1976, p. 4).

Sidorenko, V. A. et al. "Razvitiye podkhoda k resheniyu voprosov bezopasnosti atomnikh istochikov energosnobzheniya v SSSR v svyazi s rasshireniyem masshtaba i oblasti ikh primeneniya." In *Current Nuclear Power Plant Safety Issues*. Vol. 1, pp. 261–273. (Vienna: IAEA, 1981).

Sidorenko, V. A., et al. "Safety of VVER Reactors." *Soviet Atomic Energy*, Vol. 43, No. 6 (May 1978) pp. 1101–1118 (from *Atomnoya Energiya*, Vol. 43, No. 6, December 1977, pp. 449–457).

Sinev, N. M., and B. B. Baturov. *Ekonomika atomnoy energetiki*. Moscow: Energoatomizdat, 1984.

Snowball, D. J., and S. M. Macgill, "Coping with Risk: The Case of Gas Facilities in Scotland." *Environment and Planning C: Government and Policy*. Vol. 2, 1984, pp. 343–360.

Solchanyk, Roman. "More Controversy on Nuclear Energy in the Ukraine." *RFE/RL, Radio Liberty Research Reports*, RL 231/88 (June 8, 1988), p. 3.

Solchanyk, Roman. "Ukrainian Writers Protest against Nuclear Construction Site." *RFE/RL, Radio Liberty Research Reports*, RL 336/87 (August 11, 1987), pp. 1–2.

Solchanyk, Roman. "Ukrainians Send Appeal on Nuclear Energy to Party Conference." *RFE/RL, Radio Liberty Research Reports*, RL 294/88 (June 29, 1988), p. 4.

Stepanov, I. R. *Atomnaya teplofikatsiya v rayonakh severa.* Leningrad: Nauka, 1984.

Thorton, Judith. "Soviet Electric Power after Chernobyl': Economic Consequences and Options." *Soviet Economy*, Vol. 2, No. 2 (April–June 1986), pp. 131–179.

Tokarev, Yu. I. (ed.). *Yaderniye energeticheskiye ustanovki.* Moscow: Energoatomizdat, 1986.

Tretyakova, Albina, and Barry Kostinsky. "Fuel Use and Conservation in the Soviet Union: The Transportation Sector." *Gorbachev's Economic Plans*, pp. 544–565. Study paper submitted to the Joint Economic Committee, U.S. Congress, November 23, 1987. Washington D.C.: U.S. Government Printing Office, 1987.

Troitskiy, A. A. (ed.). *Energetika v SSSR, 1986–1990 godakh.* Moscow: Atomenergoizdat, 1987.

Troitskiy, A. A. "Elektroenergetika: Problemi i perspektivi." *Planovanye khozyaystvo*, No. 2, 1979, pp. 20–24.

TsSU. *Narodnoye khozyaystvo, 1922–1982.* Moscow: Finansi i Statistika, 1982.

TsSU. *Narodnoye khozyaystvo SSSR v 1980.* Moscow: Finansi i Statistika, 1981.

TsSU. *Narodnoye khozyaystvo SSSR v 1985.* Moscow: Finansi i Statistika, 1986.

TsSU. *Strana Sovetov za 50 let.* Moscow: Statistika, 1967.

Tsygankov, V. A., et. al. "Atomnaya stantsiya teplosnabzheniya dlya otdalennykh rayonov." *Teploenergetika*, No. 12 (December 1981), pp. 23–26.

Vinck, William. "Practices and Rules for Nuclear Power Stations: The Role of the Risk Concept in Assessing Acceptability." in M. Dierkes et al., pp. 107–120.

Voronitsyn, Sergei. "Concern in Tatar ASSR about Nuclear Power Station to be Built on Kama River." *RFE / RL. Radio Liberty Research Reports*, RL 222/83 (June 7, 1983), pp. 1–2.

Voronitsyn, Sergei. "Further Debate on the Safety of Nuclear Power Stations in the USSR." *RFE / RL, Radio Liberty Research Reports*, RL 350/81 (September 7, 1981), pp. 3–4.

Voronitsyn, Sergei. "How Great is Soviet Citizen's Fear of Nuclear Radiation?" *RFE / RL, Radio Liberty Research Reports*, RL 468/82 (November 22, 1982), pp. 1–2.

Warner, David, and Louis Kaiser. "Development of the USSR's Eastern Coal Basins." *Gorbachev's Economic Plans*, pp. 528–538. Study paper submitted to the Joint Economic Committee, U.S. Congress, November 23, 1987. Washington, D.C.: U.S. Government Printing Office, 1987.

Wilbanks, Thomas. "Scale and Acceptability of Nuclear Energy." In Pasqualletti and Pijawka, pp. 9–50.

Winter, John W. *Power Plant Siting*. New York, NY: Van Nostrand Reinhold Co., 1978.

Wood, William. *Nuclear Safety: Risks and Regulations*. Washington, D.C.: American Enterprise Institute for Policy Research, 1983.

Yermakov, G. V. "Nuclear Power is the Basic Trend in the Development of Future Power Engineering." *Thermal Power Engineering*, Vol. 18, No. 4 (April 1971), pp. 10–17 (from *Teploenergetika*, Vol. 18, No. 4, April 1971, pp. 6–11).

Wolpert, Julian. "Departures from the Usual Environment in Location Analysis." *Annals of the American Association of Geographers*, Vol. 60, No. 2 (1970), pp. 220–229.

Young, Katherine, (ed.). *Decision Making in the Soviet Energy Industry*. Falls Church, VA: Delphic Associates, 1986.

Yudon, Boris. "Decision-Making in the Soviet Heat and Fuel Supply Systems: Contrdictions in Consumer Supplier Relations." In Young, pp. 59–93.

Ziegler, Charles E. "Political Participation, Nationalism and Environmental Politics in the USSR." In John Massey Stewart (ed.), *The Soviet Environment: Problems, Policies and Politics*, pp. 24–39. New York: Cambridge University Press, 1992.

Zhimerin, D. G. "The Present and Future of the Soviet Power Industry." *Thermal Power Engineering*, Vol. 17, No. 3 (March 1970), pp. 4–8 (from *Teploenergetika*, Vol. 17, No. 3, March 1970, pp. 3–6).

Index

Abramov, V. G., 102n
Academy of Sciences (AN SSSR), 56,
57, 109, 162; Siberian Branch (SO
AN SSSR), 56, 79
Adamov, E., 115n
Alekshasin, P. P., 71n, 72n
Arkhangel'sk AST, 157, 169
Arkhangel'sk Oblast', 128, 169
Armenia: indepedent state of, 138,
143–144, 148; SSR, 164–165;
Armenian AES: operation, 92;
opposition to, 122, 157, 164;
shutdown, 112, 157, 164; siting of,
90, 164
AST plant design, 76, 91–93, 122, 140,
165. *See also* District heating
ATETs plant design:
discontinuation of, 108, 109, 113,
160–161; plans for, 74, 76, 91–93,
122, 140, 160–161. *See also*
Cogeneration; District heating
Atommash plant, 83
Azerbaijan AES: cancellation of, 112,
158; local party support for, 120;
opposition to 158; siting of, 90
Azerbaijan SSR, 120

Balakovo AES: cancellation of, 169;
local support for, 144; operation,
155; opposition to, 125, 157, 169;
plans for, 82, 140
Bashkir AES: cancellation of, 157,
167; opposition to, 122–123, 126,
128, 157, 167
Bashkir ASSR, 126, 128, 167
Batov, V. V., 79
Belorussia: Communist Party, 127–
128, 161, 166; Council of
Ministers, 128, 161; indepedent
state, 148; SSR, 121, 127–128, 161,
164
Belorussian AES: cancellation of,
157–158, 164; opposition to 127, 157,
164; plans for, 127.

Beloyarsk AES, 79; operation, 90, 92,
95n, 102n, 153, 177; plans for 74,
141, 177
Belyaev, A. I., 72n
Berkovich, V. M., 87
Bilibino ATETs: operation, 74, 91,
95n, 154, 177; plans for, 74, 91, 141,
177

Central planners, 53–56. *See also*
Gosplan; Ministries, USSR
Chelyabinsk Oblast', 168
Chernobyl' AES: accident, 105n, 119;
cancellation of reactors, 107, 142,
158; operation, 81, 154, 178;
opposition to, 129, 158; plans for,
81, 142
Chigirin AES: cancellation of, 128,
157, 159–160; opposition to, 121–
124, 127–128, 157, 159–160, plans
for, 90, 133n; siting of, 90, 133n,
159–160
Chung, Han-Ku, 3, 47n, 68n
CIA, 47n–48n, 68n, 70n
CIS, 138–139, 146
Coal, 30, 32–35, 40–41, 47, 48n, 138
Cogeneration, 39, 41, 50n, 73, 90–93,
109–110, 122. *See also* ATETs;
District heating
Communist Party of the Soviet Union
(CPSU), 33, 53–55, 67, 88, 119, 121,
124, 127, 146; 19th Party
Conference, 124, 163; 24th Party
Congress, 74, 76, 81; 26th Party
Congress 120; 27th Party
Congress, 120
Containment, 86–87, 93
Cook, Constance Ewing, 12, 17, 24n–
26n
Cost factors, 61–62
Council of Ministers, USSR, 54–55,
57, 66, 107, 126–129, 160–163, 168,
170–171

About the Author

Charles K. Dodd is a geographer currently living in Seattle, Washington. He holds a B. A. in Political Economy from the University of California, Berkeley, and a M. A. in Geography from the University of Washington. His research interests have been focused on energy and environmental issues in the newly independent states of the former Soviet Union and he has published several works on the subject.